MW01317788

★ CHRISTOPHER A. DURHAM ★

PHOTOGRAPHY BY
TERRI CAMPBELL

WITH A CONTRIBUTION BY
ROB WALLACE

FOLIO 28, LLC ★ CHARLOTTE, NORTH CAROLINA

FIFTY2: The My Private Brand Project.

Published by Folio28 LLC, Charlotte, NC

www.folio28.com
www.mypbrand.com

ISBN 978-0-9915220-0-2

First Edition

* * * * *

Dedicated to my wife, Laraine —
your encouragement, intelligence
and love have made this possible.

— C.D.

* * * * *

"Whatever you do, don't play it safe.
Don't do things the way they've always been done.
Don't try to fit the system.
If you do what's expected of you,
you'll never accomplish more
than others expect."

Howard Schultz
chairman and CEO, Starbucks

CONTENTS

“Here’s to the crazy ones.
The misfits. The rebels.
The troublemakers.
The round pegs in the square holes.
The ones who see things differently.
They’re not fond of rules.
And they have no respect for the status quo.
You can quote them, disagree with them,
glorify or vilify them.
About the only thing you can’t do is ignore them.
Because they change things.
They push the human race forward.
And while some may see them
as the crazy ones,we see genius.
Because the people who are crazy enough
to think they can change the world,
are the ones who do.”

Apple Inc.

ACKNOWLEDGEMENTS

First and foremost, I must thank my wife, Laraine, and daughters, Olivia and Sarah, who have been a constant source of encouragement and inspiration as well as great sample shoppers, trend spotters and photography assistants.

When I first proposed the idea of this book to Teri Campbell and his team at Teri Studios, his immediate and unwavering answer was an emphatic YES! His body of work is astonishing, and his eye for seeing brands and packaging should shape the future of brand photography for years to come. Teri would also like to thank his contributing creative team: Sherry McKown, Robert Joseph, and Scott Martin.

My sincerest gratitude to Rob Wallace for his support and encouragement, as well as the design strategy/concept work of the Wallace Church team: John Bruno, Jessica Breglio, Carolyn Edgecomb and Katie Williams.

To Kim Justen for her honesty and candor in editing and her commitment to pushing this project forward, as well as Lynn Celmer and Sharon Putman.

Special thanks to the retailers and their teams who contributed with encouragement, great brands and occasional sample procurement: Family Dollar, SuperValu, Raley's, Staples, Best Buy, Michaels, Food Lion, and Ahold.

And for all those who have contributed as supporters, thought partners and friends: Jimmy Wilson, Tonya Raynor,Greg Palese, Perry Seelert, Todd Maute, Danielle Kidney, Denise Whalen, Scott Lucas, Tammy Deboer, Don Childs, Glenn Pfeiffer, Clay Ellis, Andres Siefken, Verona Johnson, Rick Rommel, Mike Kitz, Betsy Schowachert, Tammie Hunt, Katie Kenney, Maryann Herskowitz, Barbara Glass, Melissa Smith-Hazen, Rick Rommel, Sam Mayberry, David Palmer, Scott Burglechner, Michele Oger and Lindsey Hurr.

Thanks also to Brian Sharoff, Dane Twining, Joe Azzina and Tim Simmons at the PLMA.

The trade press who have been immensely supportive of my efforts: Laura Zielinski at *Brand Packaging Magazine*, Phillip Russo at *Global Retail Brands* magazine, John Failla at *Store Brands Decisions* and the team at *PL Buyer*.

Thanks to Beth Ventura for constantly questioning and expecting greatness, as well as my friends and colleagues at Lowe's.

A special thanks to the man who put me through retail boot camp and never compromised: Dave Godfredson.

And finally to Jim Cusson, Jared Meisel and our entire team at Theory House.

Thank you all.
— Christopher Durham

"Those who cannot
change their minds
cannot change anything."

George Bernard Shaw

PREFACE

★ ★ ★ ★ ★

In late 2007, LinkedIn, the social networking site for professionals, had successfully invaded the cubicles and careers of America. Social media was exploding and joining the site swiftly progressed to the creation of a LinkedIn group focused on my passion, Private Brands.

The new group, My Private Brand, quickly grew and evolved; as LinkedIn added features, it became a vibrant social community of Private Brand leaders. In late 2008, I expanded the group to a wordpress.com blog with the same name while continuing to work a full-time job managing Private Brands. Like most new blogs, the early readership was primarily my wife and mom (who were both, I am pretty sure, lying about reading it). However, by spring, I caught a big story and broke the biggest Private Brand story of the last ten years (Walmart's Great Value redesign). Due to the story, readership quickly ballooned, and the site was off and running.

Five years ago, My Private Brand was simply a side project that allowed me to get my feet wet in social media and blogging as well as talk about a subject that I love. At the time, there were a lot of trade magazines, consultants and brokers talking about traditional private label buying, product development and manufacturing. I saw an opportunity to build a site focused on retailer-owned brands that was designed to foster innovation, encourage debate and write the next chapter of Private Brand management. Over 3,000 posts later, the site continues to grow and push retailers to build and manage great brands. Two years ago, I jumped headfirst into consulting, working with some of the world's largest retailers to manage, develop, and grow their Private Brands.

This book grew out of a conversation with a retailer at a conference. He believed private label in America was all the same: simply a value play wrapped in bad design. I found the statement disturbing and realized it honestly was not true. This book is a celebration of what I believe to be the Fifty2 best retailer-owned brands in the U.S. across all channels. They bring their positioning and business objectives to life through great design, purpose, lifestyle and innovation.

It is my hope this book will redefine not only the perception of Private Brand in America but also its potential.

Good luck!
— CD

“For me, ‘revolution’
simply means
radical change.”

Aung San Suu Kyi

BRAND MATTERS

★ ★ ★ ★ ★

This is not a book about generics, private labels, manufacturing, brokerage, sourcing, tiers, merchandising or package design. This is a book about BRANDS. Brands that just happen to be owned by retailers.

It is the story of my quest to discover the Fifty2 best retailer-owned BRANDS in the United States. A quest that led me to purchase, taste, touch and experience more than 1,000 brands from over 200 retailers from across the country. Brands that all too frequently disappointed. There are still numerous poorly executed, low quality, ugly private labels on the shelves of American retailers. Labels whose products exist as pale imitations of national brands, as tactical illustrations of price tiering and buying strategies instead of brand strategies. In some cases, the Fifty2 are the stars of a well-executed private brand portfolio. In others, they are shining lights in a portfolio of mediocrity. This book is designed to bring the best of the best to the forefront, to place them on a pedestal not as brands to be copied, but as compelling examples of brands bringing each retail owner's strategy to life.

The words brand and branding are warm, fuzzy and poorly defined. They excite designers and make retail merchants roll their eyes and simply look for the margin. Yet we all know and demonstrate the value of brand when we choose one product over another simply because of the brand. The national brand examples are easy: Coca-Cola, Starbucks, Apple, and Honda. Private Brands are there too as an integral part of our lives, shopping lists and dinner tables. We cut our grass with Craftsman lawnmowers, prepare our dinners in Williams-Sonoma cookware and feed our kids Via Roma spaghetti for dinner on Threshold plates.

SO WHAT DOES THE WORD BRAND MEAN?

Author and entrepreneur Seth Godin defines a brand as the set of expectations, memories, stories and relationships that, taken together, account for a consumer's decision to choose one product or service over another. If the consumer doesn't pay a premium, make a selection or spread the word, then no brand value exists for that consumer.

WHY DO BRANDS MATTER?

Simply put, a brand that has developed awareness, relevance and customer preference also develops value. That value is equal to the sum total of how much extra people will pay, or how often they choose one brand over another.

Although naming, logo and package designs are all important, they are only the tactical elements that work together to create a complete and fully realized brand.

Thinking about your brand is similar to the way you think about a person. Are they a collection of their body parts, hair color, eye color, education, clothes, and experiences? Or does the sum of their attributes, personality and experience combine to make something more? The essence of a brand is not the collection of the attributes (name, logo, design, etc.) but instead is that something more; it is how our customers feel about our Brand and its products.

The purpose of the name, logo, color and package design is not to create those feelings, but to remind customers of them. If their feelings about the brand are negative, those brand elements remind them of how much they dislike the brand and its products.

When a Private Brand delights its customers, value is added to the brand. When a Private Brand irritates or disappoints its customers, it diminishes the value of the brand.

The brands included in this book are the Brands that have moved beyond the 1970s generic black and white products that first defined the industry. They have moved past the late 1980s/90s "me too" mentality that created a generation of inexpensive and undifferentiated copies. They have become modern brands bringing value to their retailers and building relationships with customers.

THE BRANDS

Each of the Fifty2 brands falls into one of three categories, each of which has a unique business goal and place within the Private Brand portfolio.

Eponymous Brand: A brand that shares the name of its retail owner. It is designed to reinforce and enhance the retail brand positioning and retail brand value (Publix, Williams-Sonoma, Trader Joe's).

Sub Brand: A brand that carries the retail owner's name along with another defining name. This is often accompanied by new brand visual language and package design. The sub brand is designed to create an additive brand attribute or positioning that is not inherent to the retail brand. The addition of the sub brand should add value to the retail brand:

- Hy-Vee + "Giving Back" = Hy-Vee One Step
- Amazon + Media/E-reader = Kindle
- Safeway + Snacking = Safeway Snack Artist.

Autonomous Brand: A brand that carries a unique brand positioning that is separate from the retail brand. The best autonomous brands have unique and memorable names, visual brand language and corresponding package design and marketing, as well as a unique target customer enabling the retailer to build relationships with those customers. At its best, an autonomous brand will accrete new value to the retail brand and has the potential to become a true asset of the company (Kobalt, Tul, Simply Balanced).

BRAND TRAITS

The Fifty2 exhibit three or more of what I believe are the traits of a modern Private Brand. They have thrown off the shackles of the generic past and created ownable relationships with their customers. In most cases, customers refer to them simply as brands, and more often than not as "their" brands.

The traits include:

Business Purpose: A focus on fulfilling a specific business need that surpasses the old-school merchandising expectations of margin or penny profits. Their business goals often focus on combatting showrooming, creating differentiation and increasing traffic. However, they can fulfill any strategic purpose a retailer needs them to.

Strong Private Brands have a clear mission and purpose.

Confidence: They have a conviction and belief in their reason for being. They are brands, and they know it.

Strong Private Brands stand up and confidently declare their right to exist.

Strong Private Brands stand up and confidently declare their right to exist.

Differentiation: Great Private Brands understand they must be different; they must be more than a cheap copy of a national brand. They must own a brand positioning that consistently and reliably solves their customers' problems. Differentiation for these brands is not simply about being different but about being meaningfully and obviously different. It creates uniqueness and ownability that translates directly to the retailer and gives consumers a reason to choose the retail brand owner over their competitor.

Strong Private Brands understand and leverage their uniqueness.

Unique and Ownable Naming: A great brand name is distinctive, memorable, easy to pronounce and emotionally appealing. It is a critical element in creating a successful Private Brand. Brand names like Kindle, Ology, and Insignia successfully differentiate in crowded markets.

Strong Private Brand names are unique, ownable and emotionally appealing.

Great Design: The best Private Brands strategically integrate world-class design into every aspect of their being. It is in everything from logo to package design to in-store marketing to advertising and product design. Design is the element that ties the brand together. It brings the brand to life and creates an ownable experience.

Strong Private Brands bring their strategy to life with world class design.

Higher Purpose: Private Brands can make a difference in our lives and our customers' lives. Brands like Hy-Vee One Step and United Oil's H2O Help to Others not only give back to their communities, they create an authentic and emotionally compelling reason for consumers to choose the brand and the retailer.

Strong Private Brands have a higher purpose and an emotionally compelling reason to believe.

Commitment: The Fifty2 have become a central and enduring focus of each retailer. Private Brand is often a strategic pillar of their business, and the brand itself is often the ultimate expression of that strategic pillar. Brands like Craftsman and Kenmore have proven that consistent and sustained focus on Private Brands can create brands that become both financial assets and customer favorites.

Retailers who own strong Private Brands consistently commit to them as long-term strategic pillars.

Target Customer: The best of the Fifty2 embrace their brand positioning and leverage it to speak directly to a specific target customer. They understand that if they address the needs and solve the problems of that specific customer they can maintain a relationship and grow a profitable brand.

Strong Private Brands know their customer and create solutions to their problems.

Strategic Brand Management: Contemporary brand management at retail is a science that evolved from the traditional brand management as defined by Procter & Gamble and the now famous 1931 memo by Neil McElroy that introduced the original concept. However, many of the retailers represented by the Fifty2 now manage extensive portfolios of Private Brands strategically designed to fill different business needs and speak to different customers. They recognize that although there is a great deal to learn from old-school CPG-style brand management, the management of a portfolio of Private Brands and each individual private brand is a distinctly different discipline. It is defined as much by guaranteed shelf space and increased profitability as it is by the highly matrixed merchant-focused culture of most retailers. These retailers are creating the future of Private Brand management.

Strong Private Brands are strategically managed as financial assets of the retailer.

Investment: Retailers who view their Private Brands as assets in a strategic brand portfolio understand that an investment in that portfolio is an investment in their company value. They commit the appropriate resources in staffing, strategy, design, innovation, quality, in-store marketing, awareness and advertising.

Strong Private Brands are owned by retailers who consistently invest in their brands.

This is the age of the retailer-owned brand. Read through the Fifty2. Agree. Disagree. Argue. Take inspiration, take hope, but whatever you do, do not copy, mimic or steal. Own your own strategy. Create and manage brands that bring that strategy to life. Make your Private Brand a destination in its own right.

WHAT ABOUT MY BRAND?

★ ★ ★ ★ ★

After reviewing hundreds of Private Brands, the truth of the state of retailer-owned brands became abundantly clear. This is still an industry that loves to copy both national brands and each other. This is still an industry dominated by retailers focused on private label as a short-term margin opportunity and not a long-term brand opportunity. This is still an industry led by merchants who focus on tactical implementation and not strategic, brand-led solutions. So here are the most commons reasons a brand was not selected:

Labeling vs. Branding: Retailers who are simply slapping a name and logo on products without any regard for consumer-focused brand positioning, consistent quality, or innovation are creating labels, not brands.

Originality: Every effort was made to pick the best, most original version of a given brand positioning or strategy. If a brand is obviously a copy of another brand or trend in strategy or design, it was not included.

Generic Naming: Labels that used any combination of generic words to create an undifferentiated identity: Great, Guaranteed, Everyday, Signature, Selections, Essential, Value, etc.

Copycat Design: Labels that directly copied either a national brand or another Private Brand.

Tiering: Labels whose sole apparent business goal was to fill a price tier have not been included. Although tiering has been a valuable merchandising tactic, it is not a brand strategy.

Ugly: This one speaks for itself. More often than not, bad design is a reflection of bad or non-existent strategy and brand management, as well as a fundamental misunderstanding of the customer.

THE FIFTY2

kindle fire HDX
Emma's Kindle
Shop
Games
Apps
Books
Music
Videos
Newsstand
IRON MAN 3
Facebook
Lady Antebellum
Minecraft
Web Browser
Mail
Twitter
Showtime
Christopher's Kindle
Audiobooks
Web
Photos
Docs
kindle FreeTime
Dan Poynter's
SELF-PUBLISHING MANUAL Vol 2
How to Write, Print and Sell Your Own Book
7"
WiFi
16 GB
QUAD-CORE 2.2 GHZ
DOLBY
FRONT CAMERA

AMAZON

KINDLE

★ 01 ★

In late 2007, Seattle, Washington-based one-time bookstore, and now mega-retailer, Amazon introduced its now iconic Amazon Kindle. The portable reader debuted not as a cheap, private label copy, but instead as a fully developed brand intent on extending the Amazon relationship into the hearts of American consumers.

In the Kindle launch press release, Jeff Bezos, Amazon.com founder and CEO, spoke of both the time invested in the new brand and its laser-focused positioning. "We've been working on Kindle for more than three years. Our top design objective was for Kindle to disappear in your hands — to get out of the way — so you can enjoy your reading."

The Kindle name wisely avoids the descriptive naming clichés (e.g. Amazon Value Reader) of traditional private labels and delivers an aspirational and memorable name that means "to light a fire." It is an apt metaphor for reading, shopping, media consumption and learning. The package design further brings the brand to life with clean, uncluttered design that allows the product to be the hero; the structure feels great in hand and further reinforces the quality of the product and relevance of the brand.

The brand experience extends well beyond that first single e-reader to include the Kindle Store that now carries more than 1 million e-books, Kindle Paperwhite, Kindle reader apps for the Kindle Cloud Reader, iPhone, iPad, Mac, Windows, Android and Blackberry, an assortment of covers, cases, screen protectors and chargers, as well as the Kindle Fire HD and Kindle Fire HDX. The latest version of the Fire HDX further extends the Kindle brand with exciting new differentiating features and services like X-Ray for music, Prime instant video downloads, and the new customer service-focused Mayday button.

Kindle is perhaps the best modern example of a retailer confidently building a relevant and consumer-focused brand. According to a 2013 study by the Book Industry Study Group (BISG), almost 40% of U.S. adults who own an e-reader preferred reading e-books on a Kindle.

teri studios
Portfolio
Personal Work
Motion
Client List
Studio Space
About
Contact
Blog
Small Fries Cookbook
Our Book
46

amazon

kindle

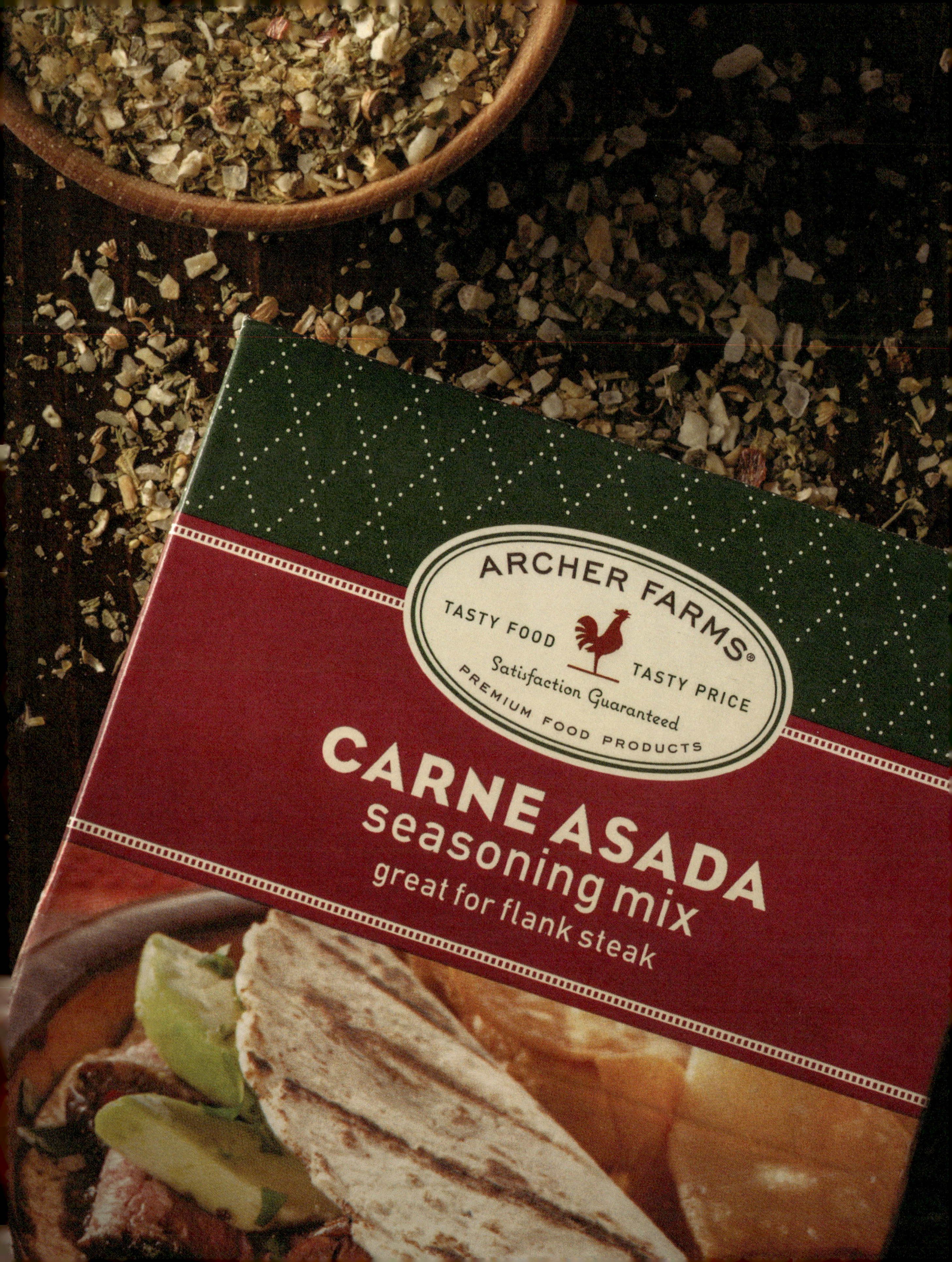
ARCHER FARMS®
TASTY FOOD
TASTY PRICE
Satisfaction Guaranteed
PREMIUM FOOD PRODUCTS
CARNE ASADA
seasoning mix
great for flank steak

ARCHER FARMS

★ 02 ★

Over the last 10 years of the modern Private Brand era, multi-tiered Private Brand portfolio architectures at American grocers have emerged as the strategy of choice. However, the strategy is typically focused on merchandising needs and the labels that emerge rarely rise to the level of brand. One rare exception is Minneapolis-based big-box retailer Target's premium and specialty foods brand, Archer Farms.

The brand eschews the derivative black-based package design and generic select/signature naming in favor of an ownable brand design anchored by an oval logo, rich green textured background, color bars in complimentary colors that simplify product selection, and professional photography that reinforces the buying decision. The name, Archer Farms, gives a playful wink and a nod to the Target Bullseye while leveraging "Farms" to create relevance in the category.

Archer Farms has consistently committed itself to bringing packaging innovation and unique flavors to Target customers. Highlights over the last few years have included the much-awarded cereal packaging and the addition of a zipper seal to potato chip bags. New tastes and flavors run the gamut from Chicken Vindaloo in box meals to Maui Onion Homestyle Kettle-Cooked Potato Chips.

In 2010, Target launched the sub brand Archer Farms Simply Balanced, which takes the guesswork out of eating right. By providing better-for-you, high-quality options that taste great, Archer Farms Simply Balanced helps Target customers make better choices about the food they eat. Simply Balanced as a sub brand evolved in 2013 to become a stand-alone lifestyle brand.

Archer Farms is the epitome of a Private Brand that has become not only an independent and compelling asset for Target, but also works to build on and reinforce their overarching retail brand positioning.

BRAND
ARCHER FARMS

BRAND FOCUS
PREMIUM & SPECIALTY FOOD

RETAILER
TARGET

HEADQUARTERS
MINNEAPOLIS, MN

ARCHER FARMS
TASTY FOOD
TASTY PRICE
Satisfaction Guaranteed
PREMIUM FOOD PRODUCTS
ROASTED
GARLIC & HERB
extra virgin olive oil
shake well
8.45 FL OZ (250mL)

ARCHER FARMS™
TASTY FOOD
TASTY PRICE
Satisfaction Guaranteed
PREMIUM FOOD PRODUCTS

ARCHER FARMS
Indian-inspired
VINDALOO
meal includes a spicy onion-ginger curry sauce, basmati rice
with spicy potato, peas sauté & mango chutney
READY IN LESS THAN 30 MINUTES
NET WT 22.6 OZ (1 LB 6.6 OZ) 643g

ARCHER FARMS®
TASTY FOOD
TASTY PRICE
Satisfaction Guaranteed
PREMIUM FOOD PRODUCTS
KOREAN-STYLE BARBECUE
slow cooker sauce
a sweet & tangy blend of soy sauce, ginger and toasted sesame seeds
ADD RIBS
refrigerate after opening
shake well
NET WT 17.1 OZ (1 LB 1.1 OZ) 485g

Artist's Loft™
Watercolor Pan Set
Ensemble de Peinture Aquarell
Juego de Moldes de Acuarela
fundamentals™ | fondamentaux | fundamentales
• Includes natural hair brush
• Wide spectrum of brilliant colors
• Water soluble
• Dissolves easily
• Inclut pinceau en poils naturels
• Vaste spectre de couleurs étincelantes
• Soluble à l'eau
• Se dissout facilement
• Incluye pincel de pelo natural
• Amplia gama de colores vivo
• Soluble en agua
• Se disuelve fácilmente

ARTIST'S LOFT

★ 03 ★

Since its introduction in 2008, Artist's Loft from Michaels Stores, Inc. has enabled customers to confidently express themselves. Whether in acrylics or oils, with pencil or pastels, on canvas or paper, in Artist's Loft, North America's largest specialty retailer of arts and crafts has the materials and supplies artists need to give their creativity form.

Michaels has created a credible, authentic and current brand in Artist's Loft that speaks directly to the artist in each of us. The exclusive brand is self-assured in its brand expression and package design, which presents itself as a brand that students, enthusiasts and professional artists know, trust and love.

Michaels has created a credible, authentic and current brand in Artist's Loft that speaks directly to the artist in each of us.

Over the last few years, Michaels has promoted the brand in-store and out with a variety of traditional and social media efforts including a sweepstakes tie-in with the 2013 film, *The Mortal Instruments: City of Bones*. In the movie, Artist's Loft products take a featured role as the preferred art supplies of the main character, Clary. The retailer also ran *The Mortal Instruments: City of Bones* Sweepstakes online promotion, hosted in-store events and created tie-in craft ideas to further engage moviegoers and artists alike.

Artist's Loft packaging is thoughtfully designed by Michaels with the artist in mind. The integrated skill-level based color-coding system on the packaging coordinates with in-store signage to simplify the shopping experience. This simple detail helps the customer feel secure about their product selection and ultimately allows the brand – and Michaels – to develop and grow a profitable relationship with artists.

Artist's Loft™ fundamentals™
Artist's Loft™ fundamentals™
Artist's Loft™ fundamentals™
Artist's Loft™ fundamentals™

Artist's Loft
Colored Pencils
Crayons de Couleur
Lápices de Color
fundamentals™ | fondamentaux | fundamentales
36 pc pz

Artist's Loft™
Drawing Pad
Bloc de Papier à Dessin
Bloc de Dibujo
fundamentals™ | fondamentaux | fundamentales
70 lb / 114 g/m²
50 sheets / feuilles /
12 in x 9 in / 30.4 cm x 22.8

Artist's Loft™
Acrylic
Acrylique
Acrílico
Academic Level
1
Niveau Académique
Nivel Académico
Ultramarine
Bleu Outremer
Ultramarino
4.06 US fl oz / 120 ml

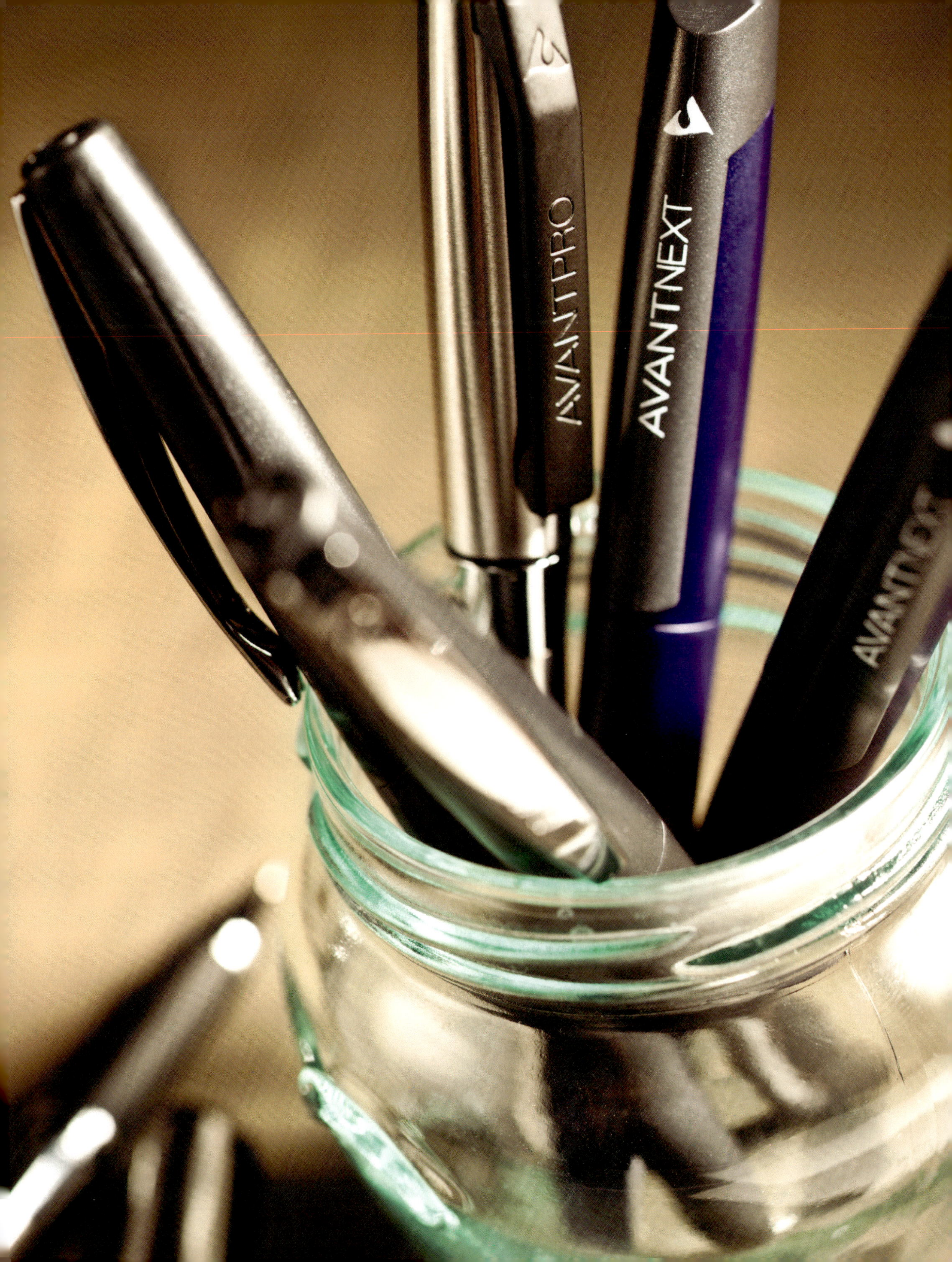
AVANTPRO
AVANTNEXT

AVANT

★ 04 ★

Since the opening of the first Staples store in Brighton, Massachusetts in 1986, the office supply retailer has built a strong portfolio of Private Brand products that consistently reinforces the Staples retail brand. Over the last year, the retailer has boldly proclaimed the Staples brand, "North America's most trusted brand in office products."

However, until 2012, the Private Brand's portfolio was primarily made up of Staples branded products, with a few sub brands and numerous product brands added to meet category needs. That all changed with the launch of their premium writing Brand, Avant. The new brand self-assuredly steps away from the Staples name and firmly establishes a credible and unique brand voice. The only hint of a Staples presence in this brand is the distribution clause.

Clean, veridical, easy-to-open packaging, as well as contemporary naming, logo and package design, reinforce the superior quality products. The minimalist product design creates ergonomic writing instruments that feel great in the hand.

The Avant brand is divided into three sub brands and color-coded by pen type for shopability: AvantNext (gel/plastic body), AvantStyle (ballpoint/plastic body), and AvantPro (ballpoint/stainless steel body). Users may customize their writing experience with interchangeable ink refills, available in black, blue and red SilkScribe and gel inks.

Avant is a promising addition to the Staples Private Brand portfolio and an intriguing taste of evolving Private Brand strategy at the world's largest office supply retailer.

AVANTPRO

AVANTSTYLE
BLACK / NOIR
1.0 mm
SilkScribe
Ink / Encre
AVANT
Gel
Ink / Encre
AVANTNEXT
BLACK / NOIR
0.5 mm
AVANT
SilkScribe
Ink / Encre
AVANTPRO
BLACK / NOIR
1.0 mm

AVANT
SilkScribe
Ink / Encre
AVANTPRO
AVANTPRO
BLACK / NOIR
1.0 mm
41657

BARA
HAND-CRAFTED
BARABOO™
WHEAT
BREWING COMPANY
Spring wheat ale with specialty malts and Yakima hops
WOODPECKER
WHEAT ALE
6–12 FL OZ BOTTLES

BARABOO

★ 05 ★

In 2012, West Des Moines, Iowa-based grocer Hy-Vee introduced the craft beer Private Brand Baraboo. To create the quality beer, named for the Baraboo River, the retailer tapped Stevens Point Brewery in Stevens Point, Wisconsin.

In the first development phase, the brand's initial three beers were created: a lager, wheat beer and an India Pale Ale (IPA). Each brew came from humble beginnings. To create just the right flavors, each label started with small, 10-gallon batches that were brewed, tested, tasted, revised, then the process repeated until the team was satisfied. Once the small-batch recipe was proven, brewmaster Gabe Hopkins had to run other samples and tests to confirm the recipe would remain consistent when brewed in larger 3,100-gallon batches.

Three distinct beer styles form the foundation of the Baraboo beer lineup. The first two are named after lumberjacks and woodpeckers, part of the Wisconsin forest culture, while the last variety name is inspired by Wisconsin's state rock, red granite. The Hy-Vee website describes the three with the brand voice:

Lumberjack IPA. Brewed with a pale malt and light hops, this one is a distinctive taste among the three beers. Though called India pale ale, this ale originated in England and was shipped overseas to British soldiers and civil servants in India during the 1800s.

Woodpecker Wheat Ale. The distinctive taste of wheat is present in this light brew. The lineage of this recipe goes back to the world's oldest brewery, which operated 1,000 years ago in Germany.

Red Granite Lager. This amber lager is in the same beer family as most popular brands. As a craft beer, however, it is made with traditional ingredients. No corners are cut for the sake of mass production, as is often the case with mass-market beers.

In addition to the three core beers, the brand has also introduced three seasonal beers, one each for the fall, winter and spring/summer.

Baraboo is a poised and bold brand with a tone of voice and graphic design that is unapologetic — this is not your grandpa's generic swill, but instead a fully developed brand and a great product.

HAND-CRAFTED
BARABOO
LAGER
BREWING COMPANY
RED GRANITE
LAGER
12 FL OZ
(355 mL)
HAND-CRAFTED
BARABOO
WHEAT
BREWING COMPANY
Spring wheat ale
with specialty malts and
Yakima hops
WOODPECKER
WHEAT ALE
HAND-CRAFTED
BARABOO
IPA
BREWING COMPANY
Crisp India pale ale,
dry hopped with Cascade, Perle
and Magnum hops
BARABOO
UMBERJACK
INDIA PALE ALE
12 FL OZ
(355 mL)

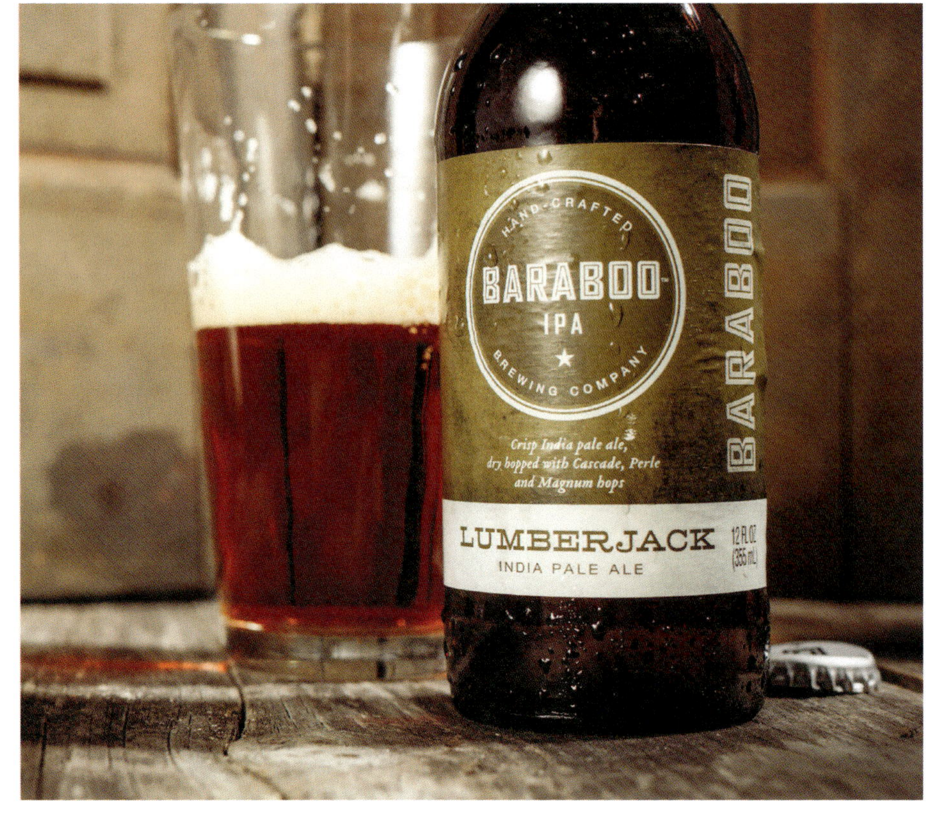
HAND-CRAFTED
BARABOO
IPA
BREWING COMPANY
Crisp India pale ale,
dry hopped with Cascade, Perle
and Magnum hops
BARABOO
LUMBERJACK
INDIA PALE ALE
12 FL OZ
(355 mL)

HAND-CRAFTED
BARABOO™
IPA
BREWING COMPANY
BARABOO
Crisp India pale ale,
dry hopped with Cascade, Perle
and Magnum hops
LUMBERJACK
INDIA PALE ALE
12 FL OZ
(355 mL)

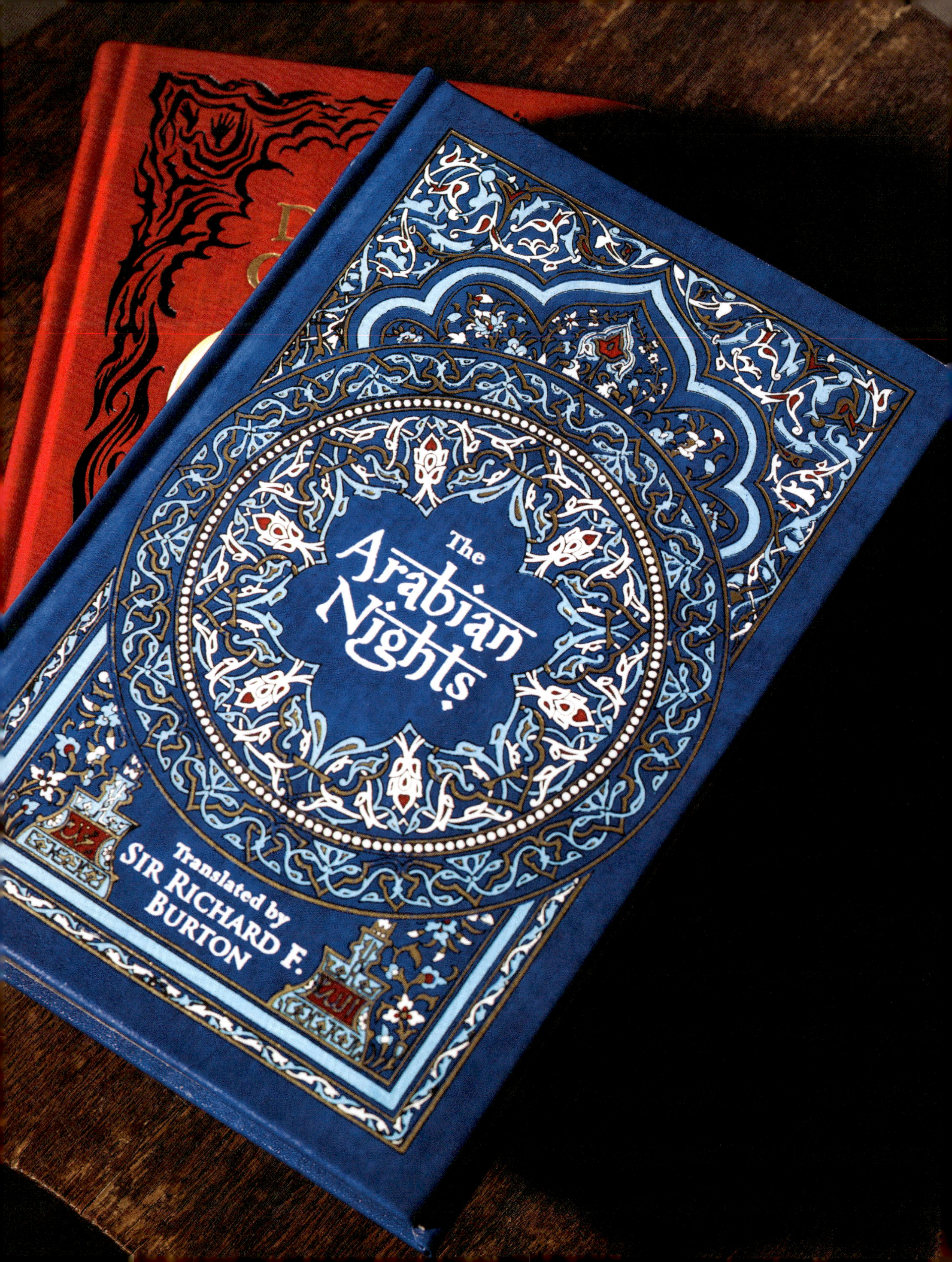
The Arabian Nights
Translated by SIR RICHARD F. BURTON

BARNES & NOBLE

CLASSICS

★ 06 ★

New York City-based bookstore Barnes & Noble began publishing its Private Brand, Barnes & Noble Classics, in early 1992. More than 20 years later, the collection has grown to more than 90 books. The leather-bound, hardcover editions include some of the greatest books ever written: *Tom Sawyer*, *Gray's Anatomy*, *Arabian Knights*, *The Complete Works of William Shakespeare*, *The Bible*, and *Alice in Wonderland*, not to mention modern classics like *The Ultimate Hitchhiker's Guide to the Galaxy*, *Dune*, *Twenty Thousand Leagues Under the Sea*, *The Star Wars Trilogy* and a bound collection of Stephen King's first three novels: *Carrie*, *Salem's Lot* and *The Shining*.

The beautifully illustrated series includes a variety of vintage illustrations and new original commissions that bring the stories to life. The books are, embossed and often feature foil stamping on the jacket and spine. These books allow book lovers to build a library that is beautiful yet sturdy, without breaking the bank. Many have marbled endpapers, painted page edges and matching ribbons and headbands, with paper heavy enough to feel significant in hand. Barnes & Noble Classics skillfully merges custom typography, illustration and well-executed layout to create beautifully bound and crafted books.

These books are the ultimate Private Brand expression of the Barnes & Noble brand. They celebrate reading, classic literature and great design to create books Barnes & Noble customers will treasure for years to come.

THE DIVINE COMEDY
DANTE
The Arabian Nights
Alice's

The Arabian Nights

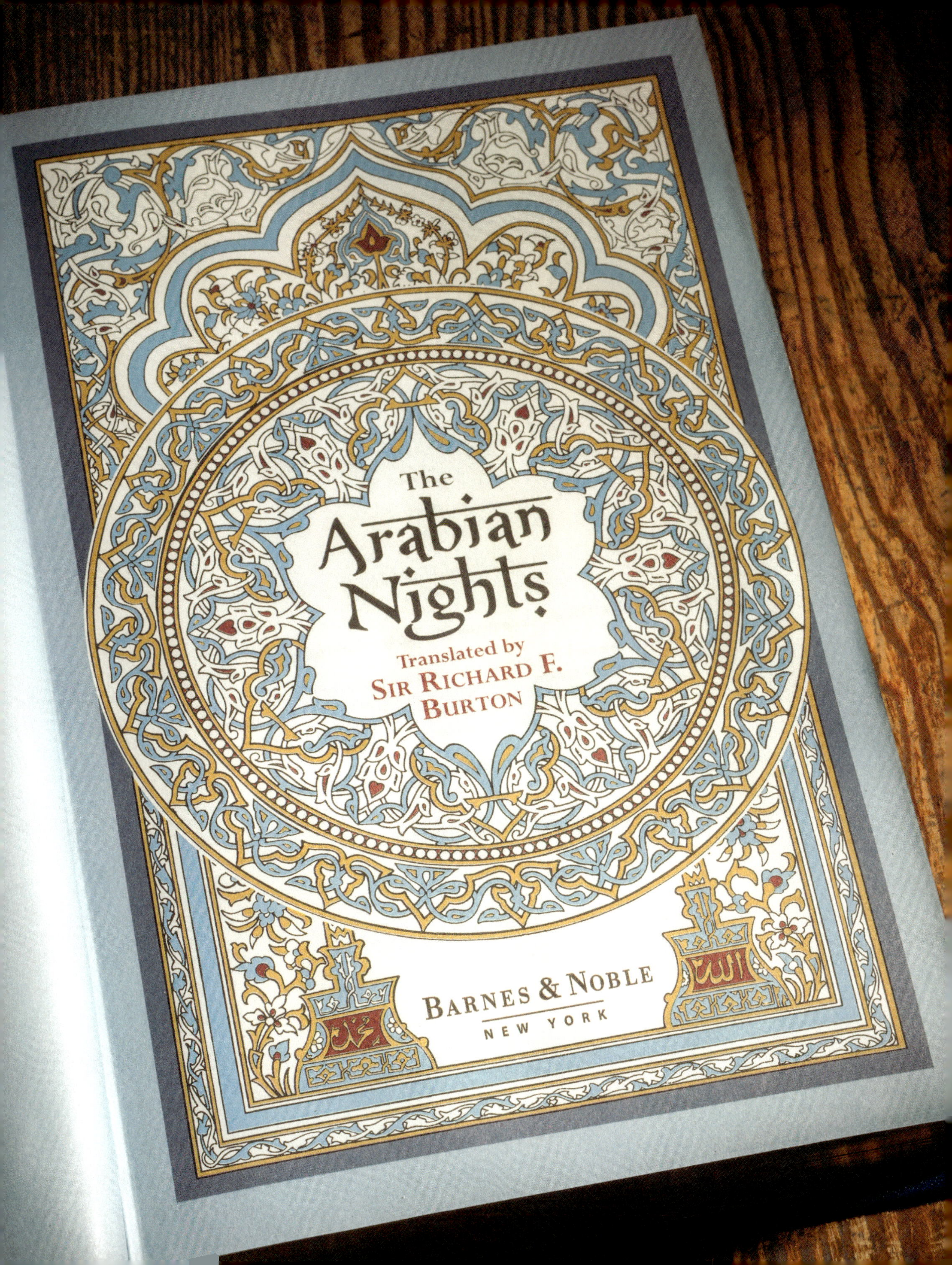
The
Arabian
Nights
Translated by
Sir Richard F.
Burton
Barnes & Noble
New York

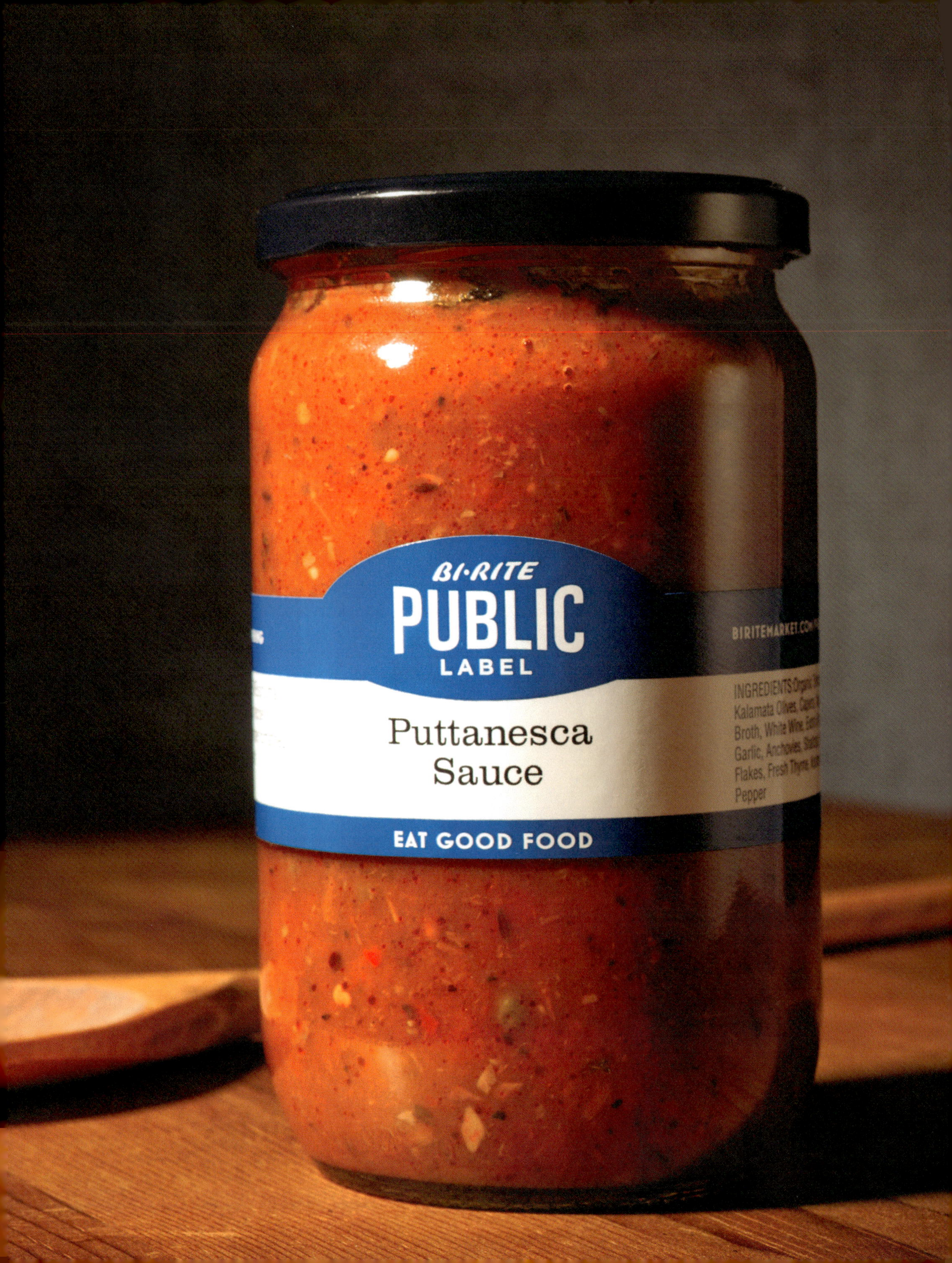
BI-RITE
PUBLIC
LABEL
Puttanesca
Sauce
EAT GOOD FOOD

BI-RITE

PUBLIC LABEL

★ 07 ★

The Mogannam family has owned the iconic San Francisco retailer Bi-Rite Market since 1964, with brothers Ned and Jack operating it for the first 26 years. Ned's sons, Sam and Raph, spent their childhood stocking the shelves, never dreaming of owning the store one day.

They took over from their father and uncle in 1997. Formerly the chef/owner of his own restaurant in downtown San Francisco, Sam brought a chef's perspective to grocery, focusing on service above all else. Today, Sam is the owner, Raph is the head grocery buyer, and Mom and Dad Mogannam still help out. The store houses a restaurant-quality kitchen and rooftop garden. The brothers also operate the Bi-Rite Creamery & Bakeshop, Bi-Rite Farms (a working farm), Bi-Rite Catering, and 18 Reasons, a community education center.

Their commitment to customer service and food are the driving forces behind the store's Private Brand. With tongue firmly in cheek, the retailer embraced the bad clichés of private label and gave its brand the ironic name Bi-Rite Public Label. They imbued it with a sense of purpose and a brand positioning that redefines what a Private Brand can be by sharing WHERE the food is from, WHO produced it and HOW it was made — and letting customers taste the products before they buy, so you know it's GOOD!

Their website confidently proclaims:

"We want to turn the 'private' in 'private label' upside down. Our line is all about transparency: we're sourcing the ingredients from farmers we have direct relationships with, partnering with kitchens in the Bay Area that have the capacity to can and jar larger quantities than we can, and providing the recipes ourselves. And we want to share the whole process with you. It's part of our constant challenge to dig deeper and learn more about how food is made, minimize food waste, and make tasty foods the old-fashioned way."

The retro-inspired brand design is restrained and uncluttered, creating a visual voice that reinforces the uniqueness of the brand and its products. Quality is paramount — many of the products are made with produce from local farms. With recipes created in the market's kitchen, many products are made in the their kitchen, while others are produced in partnership with local suppliers Happy Girl Kitchen and Community Action Marin Foodworks. In some cases, the brand uses gleaned fruits and vegetables — produce that might otherwise be left in the fields because the cost of harvest is prohibitive — purchased from farms like Mariquita near Watsonville.

Bi-Rite Public Label redefines what a private label can be and in the process creates a bold and groundbreaking BRAND.

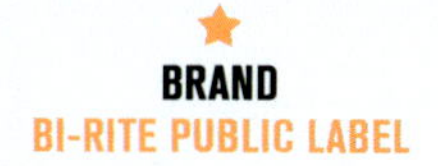

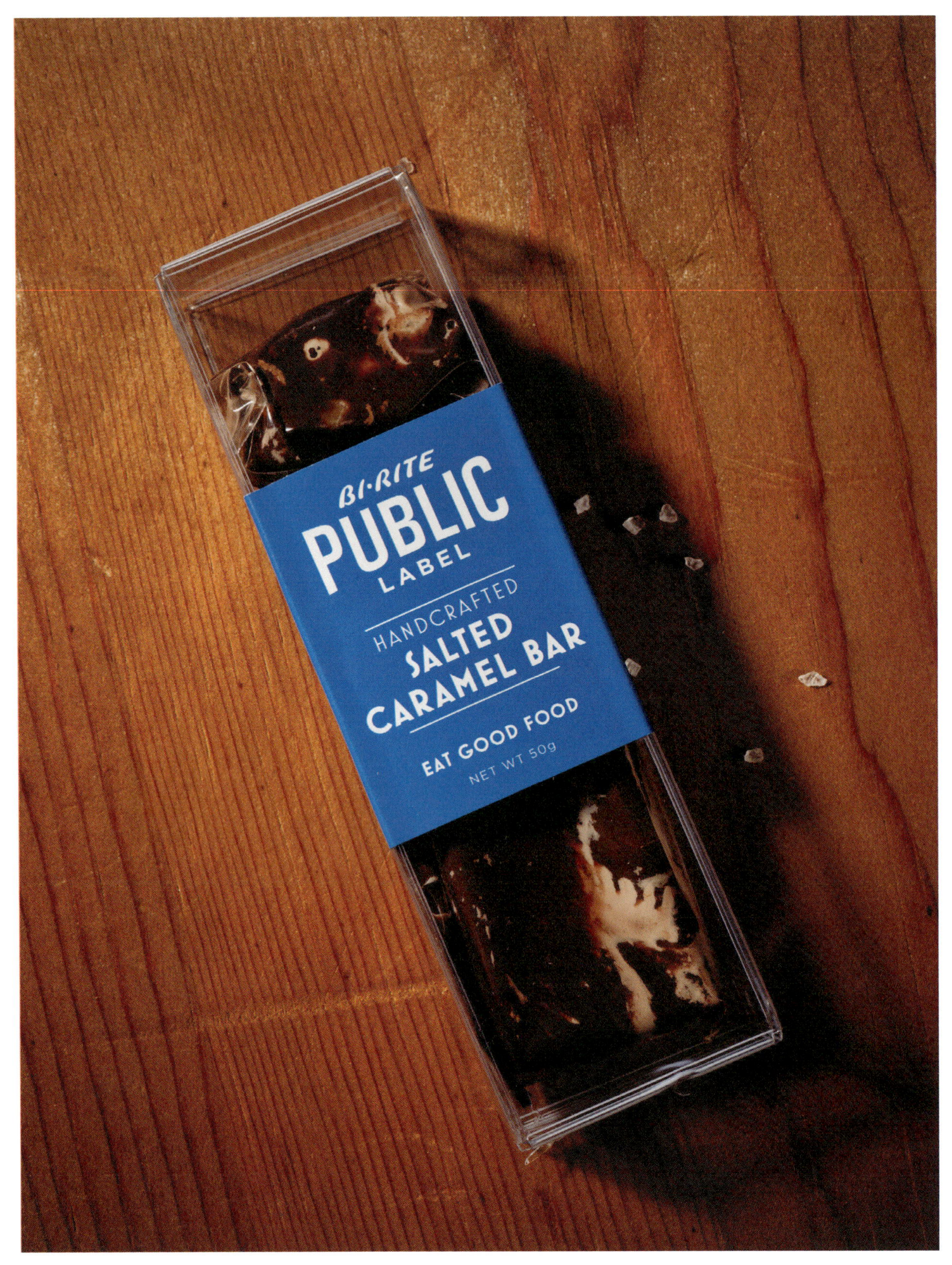
BI-RITE
PUBLIC
LABEL
HANDCRAFTED
SALTED
CARAMEL BAR
EAT GOOD FOOD
NET WT 50g

BI-RITE
PUBLIC
LABEL
French
Roast
WHO:
De La Paz
Coffee Roasters
WHERE:
San Francisco
NET WT 16 OZ

EAT GOOD FOOD
BIRITEMARKET.COM

#0386899
BLUE HAWK
DISEÑADO PARA
Pinturas Látex
5 CM
BROCHA PLANA
IDEAL PARA ÁREAS GRANDES

#0048548
BLUE HAWK
DESIGNED FOR
Oil-Based Paints & Stains
2 IN
ANGLED BRUSH IDEAL FOR TRIM AND CUTTING-IN

BLUE HAWK

★ 08 ★

Since its introduction in late 2009/early 2010, Lowe's Home Improvement's Private Brand, Blue Hawk, has expanded to more than 1,700 classic home improvement products across the store. The wide range of products includes handmade paint brushes with the iconic hawk stamped on the metal ferrule, classic leather work gloves, ropes and chains that epitomize the brand's strength, and hand and power tools that simply get the job done.

The name is brought to life by the logo with an attacking hawk that evokes tribal tattoos and impactful typography. The heroic and assertive brand design confidently defines a brand voice that combines a masculine aesthetic with utilitarian functionality.

The packaging design skillfully blends the authenticity and familiarity of the old-time American hardware store with contemporary sensibility. By presenting minimal but appropriate product attributes and benefits in an information architecture that simplifies and enhances the customer's selection process, it ultimately makes the brand engaging and easy to shop. The tan background, reminiscent of kraft paper, neutral sans serif typeface, fine-line detail and crosshatching deliver a strong, hard-wearing and professional sensibility, simultaneously evoking retro hardware package design and current lifestyle brands like Abercrombie & Fitch and American Eagle.

With its strong design, Blue Hawk sets itself apart from other traditional Private Brands and stands alone as THE BRAND with attitude.

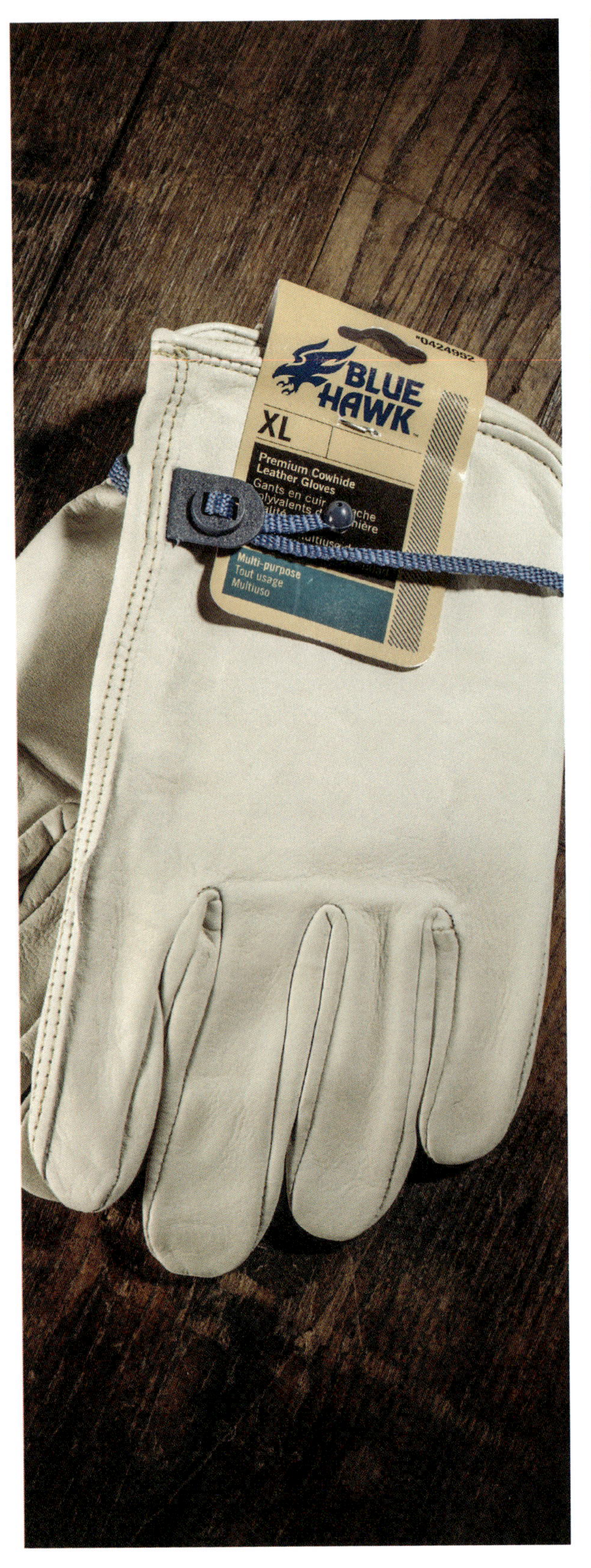
BLUE HAWK
XL
Premium Cowhide Leather Gloves
Multi-purpose
Tout usage
Multiuso

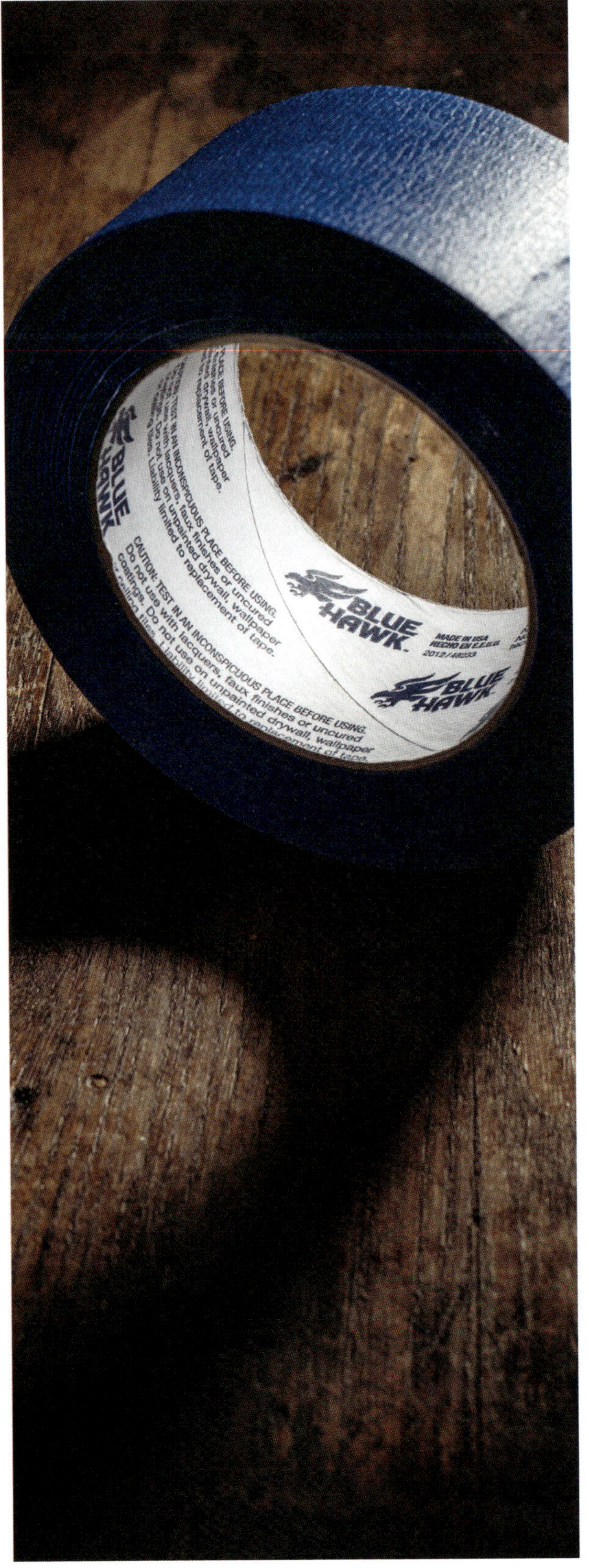
BLUE HAWK
CAUTION: TEST IN AN INCONSPICUOUS PLACE BEFORE USING.
BLUE HAWK

BLUE HAWK
SMOKE GRAY
Spray
Pintura en
resistente al
NET WT 12
340

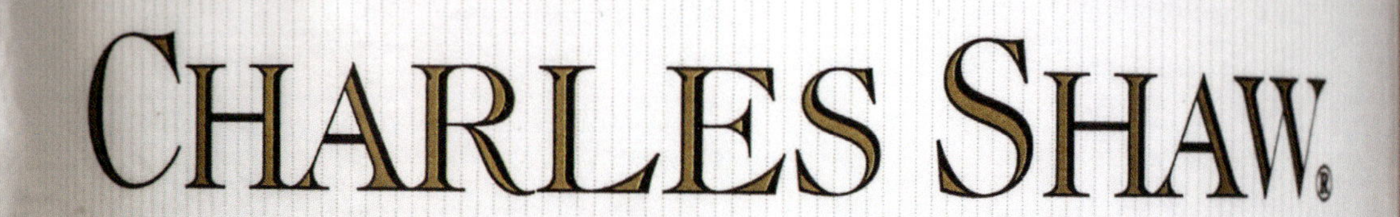
CHARLES SHAW®

2012 CALIFORNIA
Sauvignon Blanc

CHARLES SHAW

★ 09 ★

In 2012, Charles Shaw Wines celebrated its 10th anniversary as a Private Brand at Monrovia, California-based specialty grocer Trader Joe's. That's right. Ten years since the world went a bit crazy for the brand and its almost absurdly priced bottles of $1.99 wine. In that time, the retailer sold almost 600 million bottles and became one of the best known, loved and sometimes reviled wine brands in the U.S. The mixed reviews are ultimately a combination of the impact of price point, consistent quality and American wine snobbery.

Varietals of "Two Buck Chuck," as it has affectionately become known, now include chardonnay, merlot, cabernet sauvignon, shiraz, sauvignon blanc, pinot grigio, and white zinfandel. In 2009, the Charles Shaw California Pinot Grigio won a Best of Class/Gold Medal and the Charles Shaw 2009 California Shiraz won Double Gold/Best of Class and nine other medals in three wine competitions.

The 2011 Charles Shaw Cabernet Sauvignon, 2012 Charles Shaw Merlot and 2012 Charles Shaw White Zinfandel each won a Gold Medal at the 2013 Orange County Fair Wine Competition, while the 2011 Charles Shaw Chardonnay received a Bronze Medal.

The brand bears the credible sounding — and very real name — of winemaker Charles Shaw, who founded the Charles F. Shaw Winery in the late '70s and managed it through financial struggles, divorce, bankruptcy and ultimately, the sale of the brand to the Bronco Wine Company and the exclusive Private Brand distribution deal with Trader Joe's. The bottle design is simple, clean and credible, avoiding the playful and sometimes silly naming and design that has taken over wine branding in the last 10 years.

In Charles Shaw, Trader Joe's has successfully created a brand wine drinkers know and very often love; the Brand consistently gives American consumers a reason to choose Trader Joe's.

BRAND
CHARLES SHAW

BRAND FOCUS
WINE

RETAILER
TRADER JOE'S

HEADQUARTERS
MONROVIA, CA

CHARLES SHAW®
2012 CALIFORNIA
Sauvignon Blanc
Zinfandel
Merlot
CHARLES SHAW®

CHARLES SHAW
2012 CALIFORNIA
White Zinfandel
CHARLES SHAW
2012 CALIFORNIA
Sauvignon Blanc
Merlot
CHARLES SHAW
2012 CALIFORNIA

030 02 1482
circo
4 flannel receiving blankets
30 in x 30 in
(76.2 cm x 76.2 cm)

CIRCO

★ 10 ★

In 1992, Minneapolis-based retailer Target announced the arrival of a new Private Brand of clothing for infants and toddlers, featuring bright colors, color blocking and prints with pizzazz. Named “Circo,” Italian for circus, it’s available exclusively at Target.

At the time, the now infamous Ron Johnson, then merchandising manager of the Children’s Division for Target, spoke of the brand. “Circo captures the spirit of what it’s like to be a child. Childhood is a time of freedom, imagination, wonder and inspiration — and parents want to nurture this spirit in their kids. Dressing infant and toddler boys and girls in fun, bright comfortable playwear is a way to do this. Kids love bright colors.”

By the spring of 2009, the retailer was beginning a significant rethink of their Private Brand portfolio, and Circo gained importance. During the Target fourth quarter 2008 earnings release conference call, Kathryn A. Tesija, executive vice president of merchandising, said, “This spring we’ll relaunch two of our own brands [Circo & Target Home] to more clearly communicate their value to our guests. We’ve consolidated more than eight owned brands across multiple divisions to create a stronger presence for Circo, our exclusive brand for kids across all merchandise categories. This consolidation allows us to tell our guests a more complete brand story as they shop throughout the store.”

The brand then evolved from its original focus on children’s and toddler clothing to adopt a more engaging lifestyle positioning that now includes everything for kids and baby. It’s fun, colorful, well-designed packaging and products include everything from toys to apparel to bedding that both parents and kids love.

Over the last few years, *Parenting* magazine has featured Circo items including hooded towels, girls’ dresses and swimwear on its pages, and in 2012 Circo wood toys won a “Good Design Award” from The Chicago Athenaeum.

At their very best, Private Brands overcome price tiers and merchandising categories to become a trusted part of our lives. For thousands of moms across America, Circo has earned that place in their hearts and homes.

BRAND
CIRCO

BRAND FOCUS
CHILDREN’S FASHION, TOYS & ACCESSORIES

RETAILER
TARGET

HEADQUARTERS
MINNEAPOLIS, MN

circo
quarter-crew
socks

circo
washcloths
love 'n nature

8
circo®

CiRCO
8

CLARK+
KENSINGTON™
PAINT + PRIMER IN ONE
LIFETIME WARRANTY
PREMIUM INTERIOR

CLARK+
KENSINGTON
PAINT + PRIMER IN ONE
PREMIUM INTERIOR
FLAT (NON-GLARE)

CLARKE+KENSINGTON

★ II ★

In 2011, Chicago, Illinois-based Ace Hardware launched their new Private Brand of premium paint, Clark+Kensington. The line debuted with a premium line of paint and primer in one that is 100% acrylic, low-VOC paint. The name Clark+Kensington commemorates Ace's first retail location on Chicago's Clark Street and the site of their corporate headquarters.

Ace is impressively committed to maintaining the quality of the paint and the integrity of the brand by manufacturing the paint and primer in its two manufacturing facilities in the south Chicago suburbs of Matteson and Chicago Heights, versus competitors who outsource it to large paint companies.

"The Clark+Kensington paint and primer in one truly reinforces our core brand message of helping people maintain their homes while saving them time," said Mary Rice, president/general manager of Ace's paint division, in a press release at the brand launch. "The new Clark+Kensington Brand will deliver everything consumers want from their paint: fashion, style, inspiration and sophisticated colors that create inviting spaces." She went on to say, "This is an exciting moment for Ace Paint, a legacy brand in and of itself that will continue to innovate and deliver the highest-quality products to consumers looking for ways to maintain their homes quickly and easily," said Rice. "The name Clark+Kensington is a nod to our heritage, our present and our future."

In early 2012, Clark+Kensington starred in the retailer's first-ever, fully integrated campaign dedicated entirely to paint. Taking a page from national brand strategy Ace leveraged extensive consumer research to discover most consumers want and need validation and guidance in the color selection process and that they have come to believe that "do-it-yourself" means "do it alone." They recognized the color selection journey is a very personal process, an expression of a customer's own unique style.

And so out of consumer insights, a creative and highly successful television campaign, "Find Your Soul Paint" was born. In the commercial, viewers meet a woman who comes to Ace with a color in mind. As she tries to explain the color she wants, she has difficulty articulating her vision for the color; "I'm looking for a purple, but not like my favorite dress in college kind of purple… it was just a little too purple. I'm looking more for puuuuur-ple, kind of like it's raining, only it's raining way, way, way, way over there purple. Know what I mean?" To which the store associate responds: "Yeah. You want the color, but you don't want to be smothered by it. Something you can live with for a long time." The camera pans in on the perfect color brought to life in

CLARK + KENSINGTON
PAINT + PRIMER IN ONE
CLARK + KENSINGTON™
PAINT + PRIMER IN ONE

CLARK + KENSINGTON™
PAINT + PRIMER IN ONE
LIFETIME WARRANTY
PREMIUM INTERIOR
FLAT (NON-GLARE)
WITH CERAMIC MICROBEAD TECHNOLOGY
+ MATTE FINISH, HIDES IMPERFECTIONS
+ SUPERIOR COVERAGE, ULTRA SMOOTH
+ ENVIRONMENTALLY SMART, LOW VOC
946 mL (1 Qt) 32 FL OZ
124A310 Ultra White (Tint Base)

a group of handsome men in various shades of purple; this is the color she has described and there is love-at-first-sight.

The "Find Your Soul Paint" concept was then brought to life with a fully integrated strategic marketing campaign including: national television and print advertising, in-store, targeted online, social media and radio spots, as well as a complete lineup of online videos that bring the campaign's story to life.

In 2013, the hard work and commitment to the Clark+Kensington brand was rewarded when Consumer Reports named it the winner for best interior paint, beating out home improvement rivals Valspar from Lowe's and Behr from Home Depot. The magazine noted the paint was "superb at hiding, leaving a smooth finish that resisted stains and scrubbing."

In Clark+Kensington, Ace has created a memorable Brand that evokes their heritage and backed it up with a consistent and reliable quality product customers can trust.

71 36521
CRAFTSMAN
40:1 MIX
HIGH PERFORMANCE
2-CYCLE FUEL
NO ETHANOL/ALCOHOL
Mezcla de combustible 40:1 2-ciclos
Convenient Time-Saving Solution Eliminates the Hassle and Guesswork of Mixing Oil and Gas to the Correct Ratio
Ready-To-Use 40:1 Fuel Mix for Yard Tools with Air-Cooled Engines Such as String Trimmers, Blowers and Chain Saws
Specially Formulated with Full Synthetic Oil, Fuel Stabilizer and Ethanol-Free Fuel

71 10582
CRAFTSMAN
50:1 MIX
HIGH PERFORMANCE
2-CYCLE FUEL
NO ETHANOL/ALCOHOL

EXTREMELY FLAMMABLE LIQUID
DANGER:
EXTREMELY FLAMMABLE LIQUID & VAPORS
HARMFUL VAPORS
HARMFUL OR FATAL IF SWALLOWED
VAPORS MAY CAUSE FLASH FIRE
CAREFULLY READ ALL WARNINGS ON FRONT AND BACK PANELS BEFORE USE
KEEP OUT OF REACH OF CHILDREN
EMERGENCY TREATMENT:
HARMFUL OR FATAL IF SWALLOWED. DO NOT INDUCE VOMITING.
Give milk or water. Call physician immediately. Contains gasoline.
PELIGRO:
LÍQUIDO Y VAPORES EXTREMADAMENTE FLAMABLES
VAPORES NOCIVOS
DAÑINO O MORTAL SI ES INGERIDO
LOS VAPORES PUEDEN CAUSAR FUEGO
LEA CUIDADOSAMENTE TODAS LAS ADVERTENCIAS EN PANELES DELANTEROS Y TRASEROS ANTES DE USAR
MANTENGA FUERA DEL ALCANCE LOS NIÑOS
TRATAMIENTO DE EMERGENCIA:
DAÑINO O MORTAL SI ES INGERIDO. NO INDUZCA EL VÓMITO.
Ingiera leche o agua. Llame a un doctor de inmediato. Contiene gasolina.
32 FL. OZ. (946 mL)

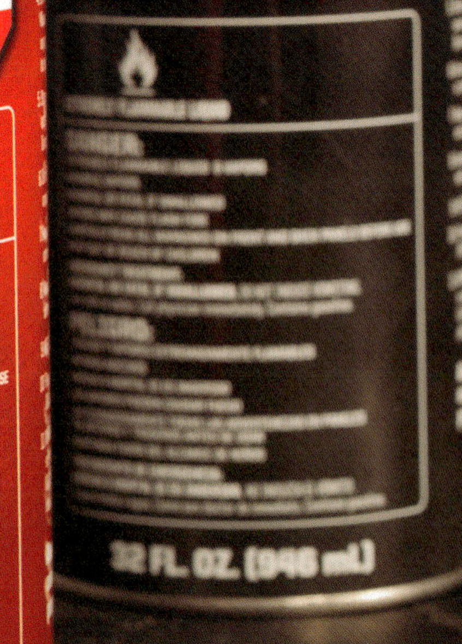

32 FL. OZ. (946 mL)

CRAFTSMAN

★ 12 ★

There are few brands in America — much less Private Brands — that inspire loyalty and passion like Craftsman Tools. In 2012, the brand was estimated to have more than $2.5 billion in sales with over 6,000 SKUs in 80 different business segments, with a 33% market share in the hand tool market and a 14% share in the power tool market. Over the last few years, Craftsman has broken out of the Sears toolbox and been introduced in Kmart, Orchard Supply Hardware, Ace Hardware, US Military Army and Air Force Exchanges, Costco, Grainger and Summit Racing Equipment.

In 1927, Sears hired hardware expert Arthur Barrows to lead the hardware department. Barrows believed the retailer needed a credible, ownable brand that differentiated them from manufacturers. He liked the name Craftsman (at the time owned by the Marion-Craftsman Tool Company), and Sears acquired the Craftsman trademark in October of that year for $500. That same year, Sears introduced the first line of Craftsman branded hand tools complete with the now legendary unconditional lifetime warranty in the Sears hardware and cutlery catalog.

After 86 years, the brand continues to grow and innovate — buttressing their ailing retail parent and consistently winning consumer trust and quality awards. Craftsman was singled out as America's top power tool brand by the 2011 Harris Poll EquiTrend study. In 2010, *Popular Mechanics* magazine awarded Craftsman the 2010 Reader's Choice Award and named the brand the favorite hand tool brand for the second year in a row.

Craftsman has moved well beyond your grandfather's garage to become a brand experience. In 2010, the retailer opened the Craftsman Experience store in Chicago, completely immersing customers in the brand and featuring a variety of hands-on activities, learning opportunities and product testing.

In 2012, they leveraged the excitement of the U.S. presidential race to create an innovative promotional

BRAND
CRAFTSMAN

BRAND FOCUS
TOOLS

RETAILER
SEARS HOLDINGS

HEADQUARTERS
HOFFMAN ESTATES, IL

CRAFTSMAN
CORDLESS SCREWDRIVER
DESTORNILLADOR INALAMBRICO
3.6 VOLT
11139
CRAFTSMAN
BriteDriver
3.6V
1 YEAR WARRANTY

5/8
14018
CRAFTSMAN
7-PC. UNIVERSAL WRENCH SET
JUEGO DE LLAVE UNIVERSAL DE 7 PIEZAS
Universal Design
One Wrench, Many Jobs
11/16
12pt
Hex (6pt)
Partially Rounded Hex
Square (4pt)
CRAFTSMAN

and philanthropic campaign, the House United Program. They set up outside the Republican and Democratic National Conventions. One half of a house was built by delegates and volunteers at the RNC, and the other half was built by delegates and volunteers at the DNC. The divided house was completed in Charlotte, N.C., and then donated to a military veteran. Tapping into the philanthropic nature of the event and showcasing it at the two conventions, Craftsman was able to capture the hearts of their customers.

Most recently, Craftsman attended New York Comic Con with its own custom publication, a DC comic book starring the new superhero, The Technician and his Craftsman Bolt-On tool, bringing Craftsman to the next generation of tool enthusiasts.

The current expression of the Brand in-store and on-pack both skillfully balances the equities and heritage of the last 86 years with a trustworthy, credible expression of tough. Craftsman is the living proof that retailers can create, build and manage Brands as assets that customers know and love.

Daily Chef
Whole Roasted
Red Peppers
HAND PEELED
IMPORTED PIQUILLO PEPPERS
SLOW ROASTED OVER OPEN FIRES
NET WT 33.5 OZ
(2 LB 1.5 OZ) (950g)

DAILY CHEF

★ 13 ★

In late 2011, Bentonville, Arkansas-based Sam's Club, a division of Walmart Stores, Inc., announced the introduction of three new Private Brands: Simply Right, Artisan Fresh, and Daily Chef. The announcement signaled a significant shift in the Private Brand portfolio strategy of the retailer and a movement away from the aging but predominant private label Member's Mark. Following the playbook of traditional consumer packaged goods (CPG) brands, the retailer conducted extensive consumer research to determine what consumers valued most when shopping. After more than 18 months of analysis and listening, Sam's Club established three brands responding to their members' priority needs in health and wellness, fresh, flavorful meal solutions and everyday cooking.

Daily Chef, the otherwise traditional national brand equivalent (NBE) Private Brand, is built on the aspirational premise that customers' favorite foods should be delicious and affordable enough to enjoy every day. Daily Chef features carefully chosen, flavorful quality foods and ingredients in a wide range of products, such as colossal easy-peel shrimp, sea salt kettle chips, organic whole bean coffee and handmade lump meat crab cakes. That premise is then brought to life through stylish and relevant package design that is a fresh take on classic restaurant branding.

Sam's Club has continued to showcase Daily Chef and the other new brands with in-store programs offering its members the opportunity to experience the quality and consistency of these brands. In addition to the in-store activation, the brand is supported with a presence on Facebook, Twitter and Pinterest, as well as mommy blogger outreach using the Collective Bias platform.

Daily Chef is part of the new generation of Private Brands intent on building relationships with America's consumers that will surpass the traditional transactional price-focused private label paradigm and replace it with a differentiated brand experience.

Daily Chef
100% ORGANIC
Extra Virgin
Olive Oil
FIRST COLD PRESSING
PRODUCT OF
ITALY
GOURMET OIL
RICH FLAVOR
MAXIMUM 0.5% ACIDITY
USDA
ORGANIC

GOURMET FOODS

Daily Chef

Tomato
Sauce

ALL NATURAL · FAT FREE

NET WT 15 OZ (425g)

DEAN & DELUCA
Vermont Fancy
MAPLE SYRUP
100% pure, all natural, Grade A
fancy maple syrup.
NET WT 16 FL OZ (473mL)
DEAN & DELUCA
Rib Rub
NET WT 5.50 oz (155 g)

DEAN & DELUCA

★ 14 ★

The original Dean & Deluca store opened for business in September 1977 in SoHo, the edgy artist and warehouse district of New York City. Founders Joel Dean, Giorgio DeLuca and Jack Ceglic quickly grew the single store into the iconic multi-channel retailer of gourmet and specialty foods, premium wines and high-end kitchenware with operations throughout the United States and abroad. Dean & DeLuca Private Brand products are available in its 14 U.S. stores and cafés, catalogs, and on its website. The retailer also sells Private Brand products to retailers and wholesalers throughout the world.

The Dean & DeLuca brand philosophy is to let the food be the star. The Private Brand packaging reinforces the master brand and its philosophy by using clean lines, genuine materials and no gimmicks. Whether it's in the store or online, the food is the star with the packaging and brand design created to accentuate the product.

The neo-classical, minimalist design focuses on two of the Dean & Deluca core brand colors, white and silver, with a secondary color implemented to create a subtle differentiator for identifying flavors and varieties. Dean & Deluca commissioned the script font in the '70s, based on the handwriting of one of their founders.

Over its history, Dean & DeLuca has earned a reputation as an icon of gourmet/foodie culture and lifestyle. Revered as a hub for the discovery of new culinary trends and little-known culinary traditions, it's only fitting that the Dean & Deluca Private Brand package design and premium products bring that tradition to life with the same confidence and elegance of the in-store experience. This is a Private Brand that perfectly complements its master brand and brings that brand to life in home and at the dinner table.

Pappardelle
Rigate
PASTA
Toscano
DIPPING OIL
DEAN & DELUCA
Vermont Fancy
MAPLE SYRUP
100% pure, all natural, Grade A
fancy maple syrup.
NET WT 16 FL OZ (473mL)
Rib Rub

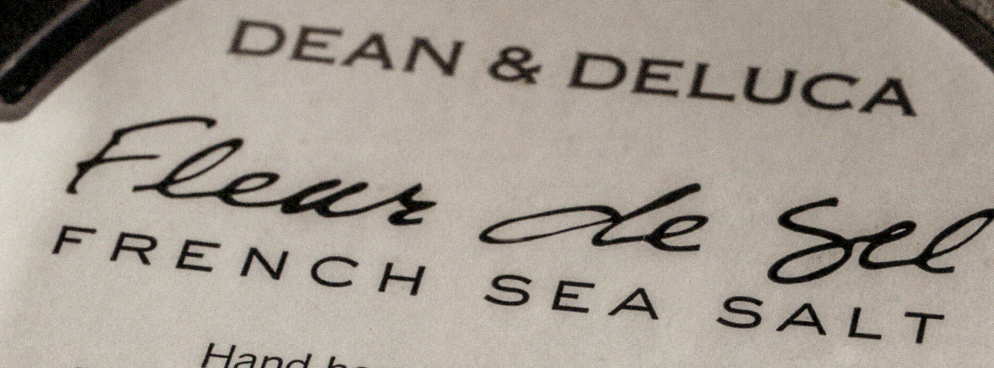
DEAN & DELUCA
Fleur de Sel
FRENCH SEA SALT
Hand harvested in France.
Perfect for salads, vegetables, seafood,
chicken, red meat and finishing.
NET WT 8 OZ (228g)

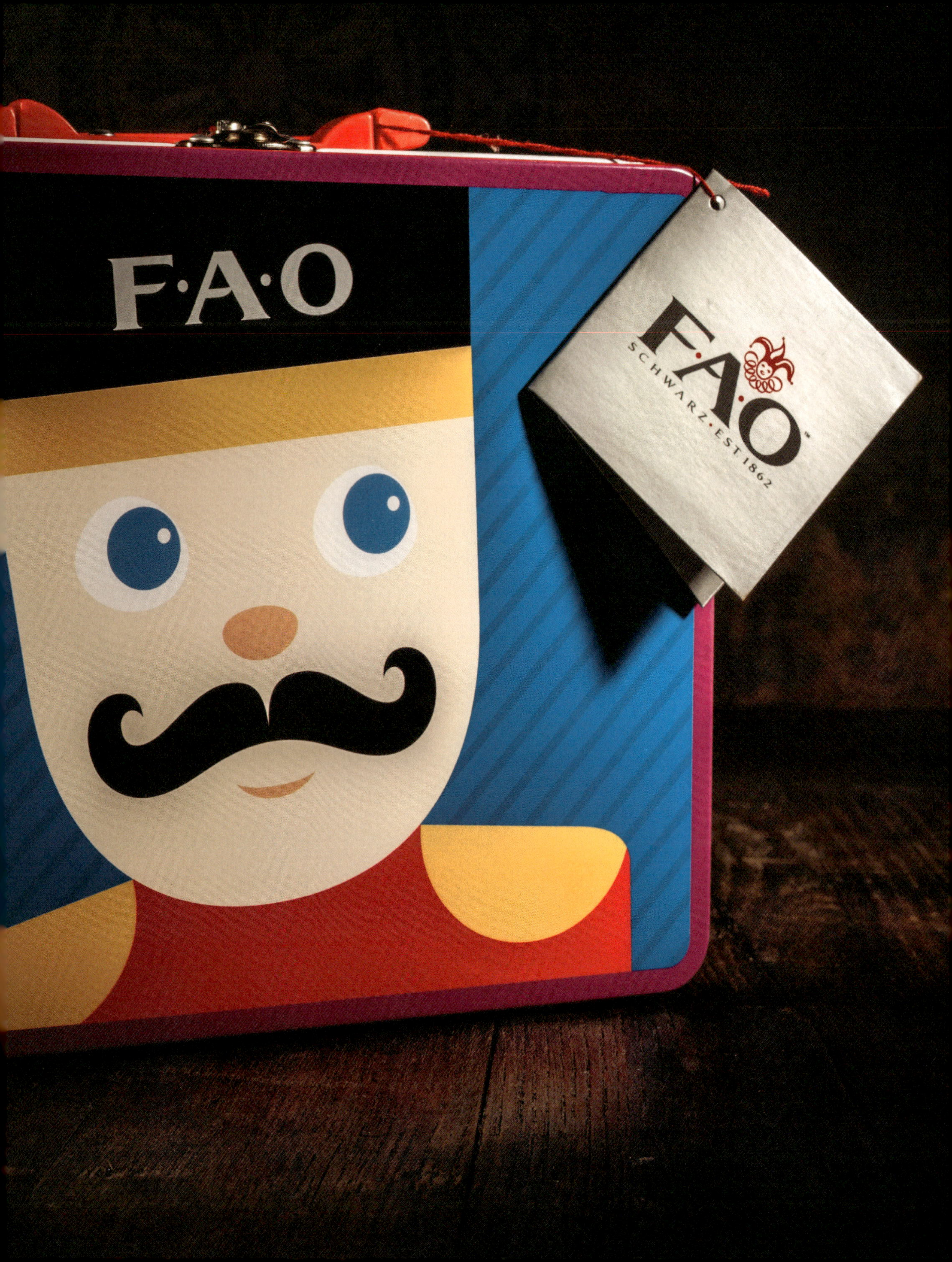
F·A·O
F·A·O
SCHWARZ·EST 1862

FAO SCHWARTZ

★ 15 ★

More than 150 years ago, a German immigrant, Frederick August Otto Schwarz, had a vision to establish a magical toy emporium featuring extraordinary, one-of-a-kind toys from around the world. He would present them in a store, showcasing the toys in a theatrical retail experience that delighted children and brought the toys to life.

In 1856, Frederick, the youngest of three Schwarz brothers, emigrated to the U.S. from Germany, joining his brothers in Baltimore, Maryland. Six years later, the brothers opened Toy Bazaar, their first toy retailing business. Following the success of that store, Frederick moved to New York City in 1870 where he opened Schwarz Brothers — Importers. Driven by his love of toys, Frederick stocked his store with playthings from Europe. By 1900, Frederick had renamed the store FAO Schwarz, and many considered him to be the largest toy dealer in the world.

A pop culture icon for generations, Frederick's flagship Fifth Avenue store has become a must-see tourist destination. Over the years, FAO Schwarz has starred in countless movies — including the much loved *BIG* in 1988, in which Tom Hanks and Robert Loggia dance on the store's giant foot piano. Most recently, the store was featured in the 2011 live-action, 3D film dedicated to The Smurfs.

In 2009, FAO Schwarz was acquired by Toys"R"Us, Inc. Over time, the retailer evolved from a simple toy store to an iconic toy brand, which, after its acquisition, is now available not only in FAO Schwarz stores, but in Toys"R" Us stores nationwide. The premium Private Brand now encompasses more than 300 classic playthings that bring the historic aspirations of founder Fredrick to life including the *BIG* (Floor) piano, classic red derby pedal car, princess jack in the box, children's clothing and the signature line of oversized plush toys that can be seen in the flagship store and experienced at home in smaller, more-affordable versions. Each is merchandised in packaging designed to evoke the heritage of FAO Schwarz while maintaining a fresh, modern sensibility.

FAO Schwarz as a Private Brand is the perfect extension of their toy retailing heritage and a beloved part of American childhood.

BRAND
FAO SCHWARZ

BRAND FOCUS
CLASSIC TOYS

RETAILER
TOYS "R" US

HEADQUARTERS
WAYNE, NJ

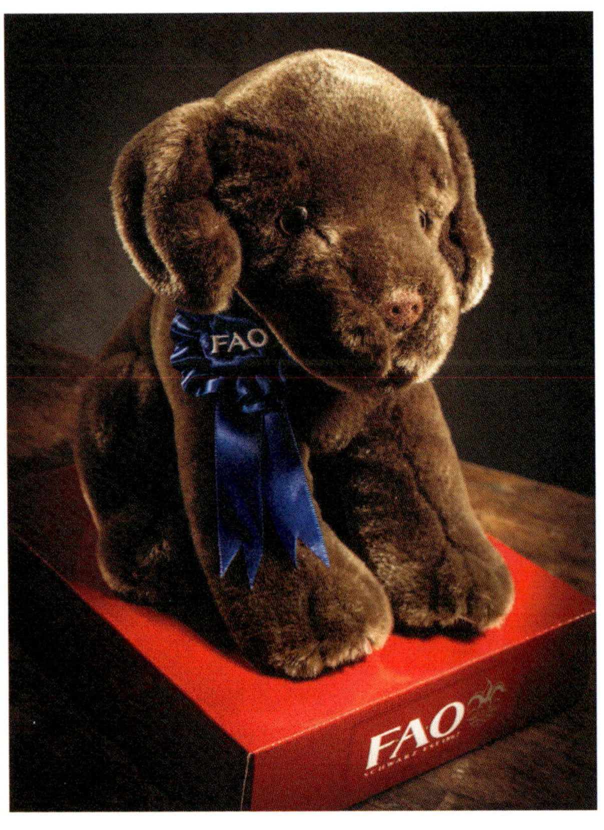
FAO
FAO

F·A·O

FAO EXPRESS

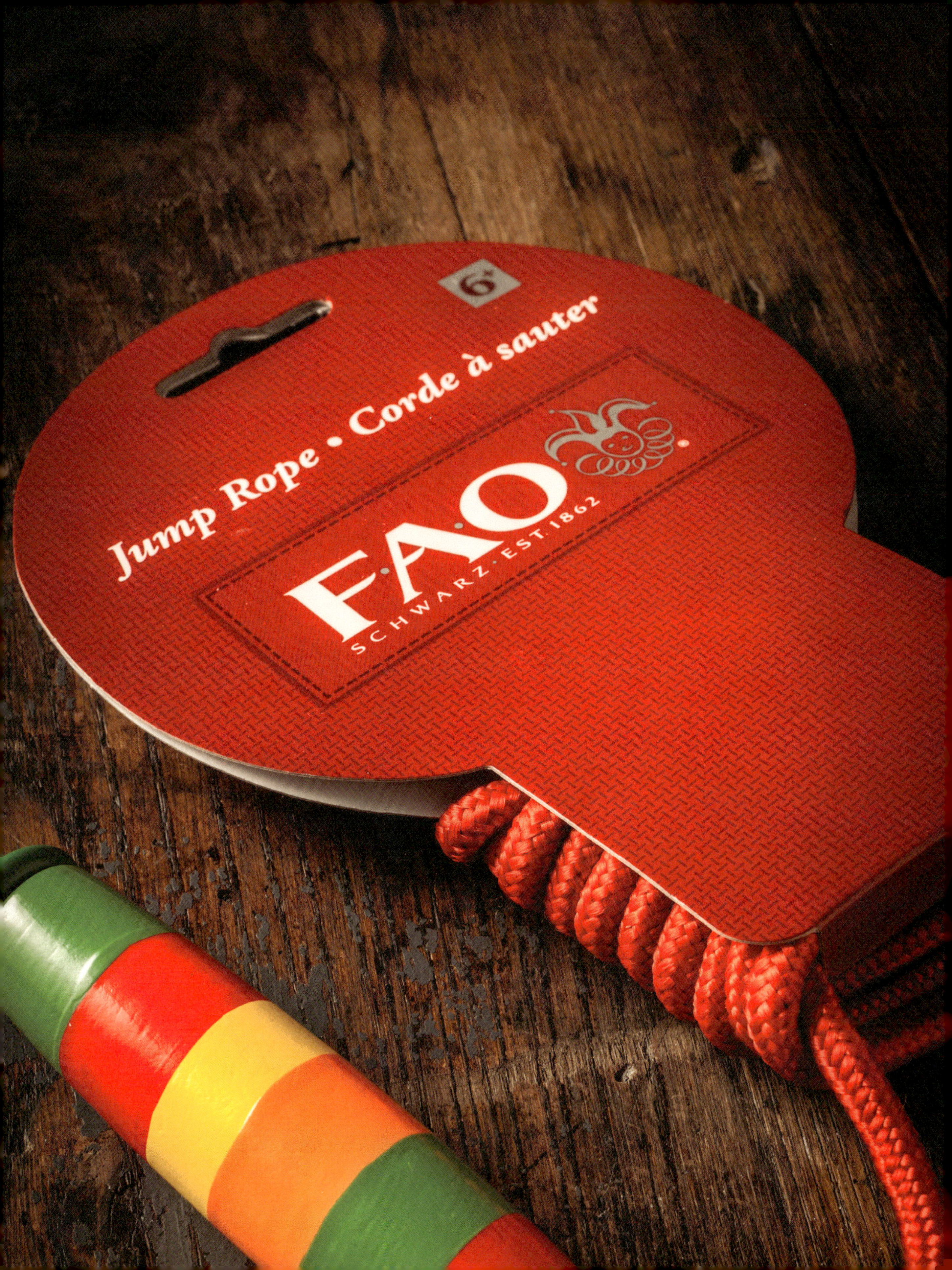
6'
Jump Rope • Corde à sauter
F·A·O·
SCHWARZ · EST. 1862

fresh & easy
MAPLE LEAF
Crèmes
tasty shortbread cookies, stuffed with luscious Canadian maple syrup crème
PER 1 COOKIE
110
NET WT. 11.4 OZ (323g)
fresh & easy
GINGER LEMON
Crèmes
tasty ginger cookies filled with
not actual size serving suggestion
25mg SODIUM
10g SUGARS
fresh & easy
CHOCOLATE CHIP FILLOOKIE™
fresh & easy
CHOCOLATE CHIP
FILLOOKIE™
2
fully-baked, soft, buttery chocolate chip cookie, with a luscious, semi-sweet chocolate crème center
NO ARTIFICIAL FLAVORS, COLORS OR PRESERVATIVES
Perishable: Keep Frozen
NET WT. 8 OZ (228g)
serving suggestion
individually wrapped
PER 1 COOKIE
500 CALORIES
11g SAT FAT 55% DV
340mg SODIUM 14% DV
43g SUGARS

FRESH & EASY

★ 16 ★

Over the last five years, British mega-retailer Tesco's grand U.S. experiment, Fresh & Easy, has transformed from being the next great concept in American retail to a beleaguered chain for sale to the highest bidder. However, since its very inception, Private Brand has been a strategic emphasis at the almost 200 stores in California, Nevada and Arizona.

The retailer launched with more than 1,000 all-natural Private Brand products and within four years the SKU count exceeded 3,000. Each Private Brand product was tasked with living up to the quality of the Fresh & Easy brand positioning and clearly bringing it to life at their customers' dinner table. Consequently, all Fresh & Easy Private Brand products adhere to comprehensive food safety standards and strict brand standards that prohibit artificial flavors, colors, and preservatives, as well as no high-fructose corn syrup or trans fats in any of the products.

A new breed of Private Brand that embraces an adventurous, playful and vibrant American spirit

Despite some early missteps that brought an unfamiliar British aesthetic to the shelves, the in-store experience and Private Brand design have continued to evolve and embrace the American customer. Consequently, a new breed of U.S. Private Brand design has emerged that not only owes its roots to its Tesco parentage, but also embraces an adventurous, playful and vibrant American spirit. The brand has received more than two dozen awards for graphic design and innovation from some of the world's most prestigious contests, including the Pentawards, the American Graphic Design & Advertising Awards, and the DBA Design Effectiveness Awards, as well as awards hosted by *Graphic Design USA*, *PLBuyer*, *Brand Packaging Magazine*, *The Dieline* and *How Design Magazine*.

No matter what happens to Fresh & Easy, their Private Brand is a bold experiment that has raised the bar for years to come.

In mid 2013, Fresh & Easy was sold by Tesco to American investors.

fresh & easy
HIMALAYAN PINK SALT
fresh & easy
MEDITERRANEAN
SEA SALT
from southern Italy – superior quality, pure crystals, mild taste
NET WT. 4.6 OZ (132g)
fresh & easy

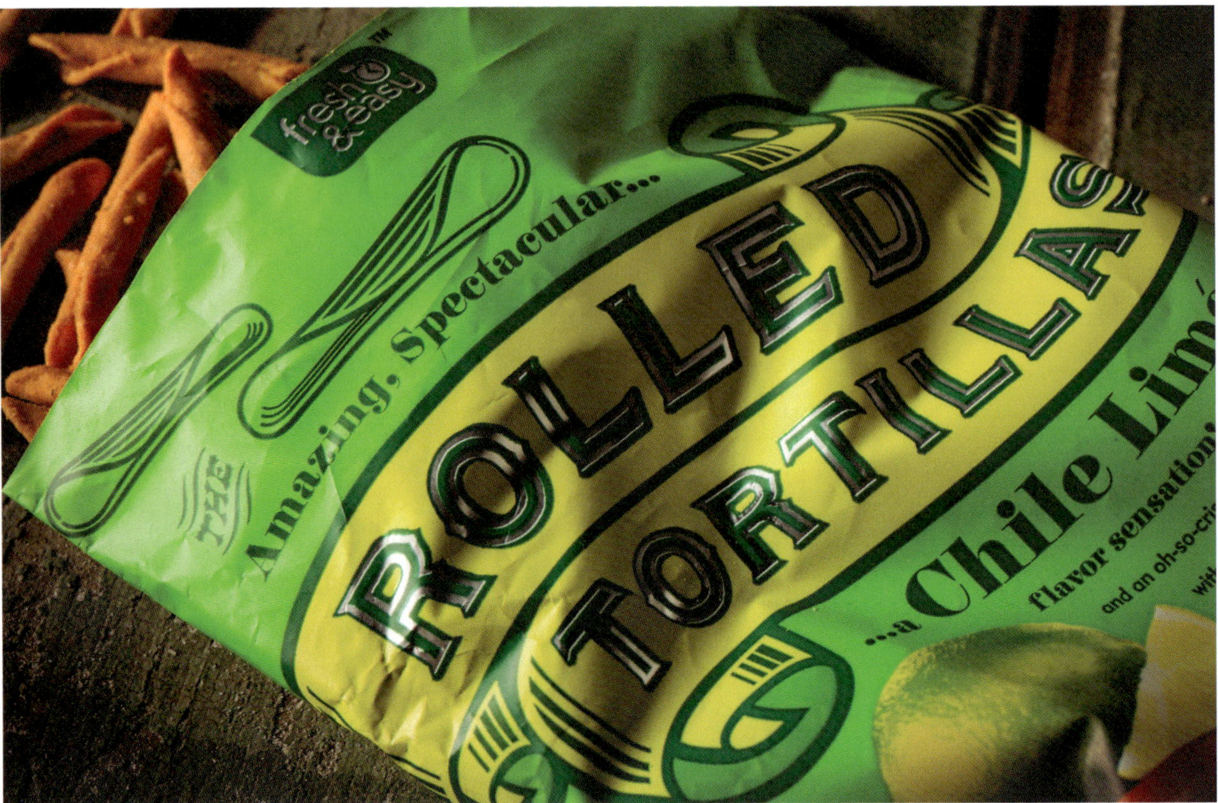
fresh & easy
THE
Amazing, Spectacular...
ROLLED
TORTILLAS

fresh & easy™ cookie bites
fresh & easy™ apple & cinnamon smiles
fresh & easy™
apple & cinnamon smiles
Our apple & cinnamon smiles cereal will make your kids smile with the great taste of apple juice and cinnamon in a crunchy treat.
serving suggestion enlarged to show detail
NO ARTIFICIAL PRESERVATIVES FLAVORS OR COLORS
EXCELLENT SOURCE OF CALCIUM & VIT D
NET WT. 11 OZ (311.8g)

gold
emblem
absolutely divine
oatmeal raisin cookies
0g trans fat per serving
see nutrition information for saturated fat content
serving suggestion
enlarged to show texture
per 2 cookies
120 CALORIES
2.5g SAT FAT 13% DV
115mg SODIUM 5% DV
10g SUGARS
NET WT
10.6 OZ (300 g)
chip cookies
te chips • 0g trans fat per serving
see nutrition information for saturated fat content
serving suggestion
enlarged to show texture
per 2 cookies
150 CALORIES
4.5g SAT FAT 23% DV
85mg SODIUM 4% DV
9g SUGARS

GOLD EMBLEM

★ 17 ★

Woonsocket, Rhode Island-based drug retailer CVS originally registered their food and consumables brand, Gold Emblem, in 1995. On Thursday, December 13, 2012, the retailer announced the redesign of the 260 Gold Emblem SKUs at its annual Analyst Day in New York City. The much-needed redesign and refocus on quality in late 2012/early 2013 revived the aging brand.

At the event, Mark S. Cosby, executive vice president and president of CVS Pharmacy, provided a closer look at the evolution and optimization of the Private Brand portfolio as a core differentiating retail growth strategy.

"In 2013, we are redefining our Gold Emblem consumable brand. The brand has been rebuilt to provide improved taste appeal and a new contemporary packaging design. We will update the packaging in all 260 of our SKUs, and we'll improve the product quality in 25% of those SKUs based on taste and quality test assessments. The brand will be relaunched in stores in January, just in time for the Super Bowl."

The redesign of Gold Emblem is a fresh take on the traditional national brand equivalent Private Brand. The brand voice and identity is shifting from dated and old to organized and sure, separating itself from the CVS identity to stand alone as a brand. The brand architecture is expanding to include two tiers: Gold Emblem and Gold Emblem Select (stay tuned; no Select was on shelf at this date).

The redesign of Gold Emblem enables the CVS Private Brand portfolio to compete more effectively in food categories.

BRAND
GOLD EMBLEM

BRAND FOCUS
MAINSTREAM

RETAILER
CVS

HEADQUARTERS
WOONSOCKET, RI

gold
emblem
peanut butter blend trail
peanuts, honey roasted peanuts, M&M'S® milk chocolate
candies & Reese's® Pieces® peanut butter candies
blend trail mix
raisins, peanuts, M&M'S® milk chocolate candies,
almonds & cashews
gold
emblem
cajun blend trail
hot cajun corn sticks, pretzels,
sesame sticks & almonds

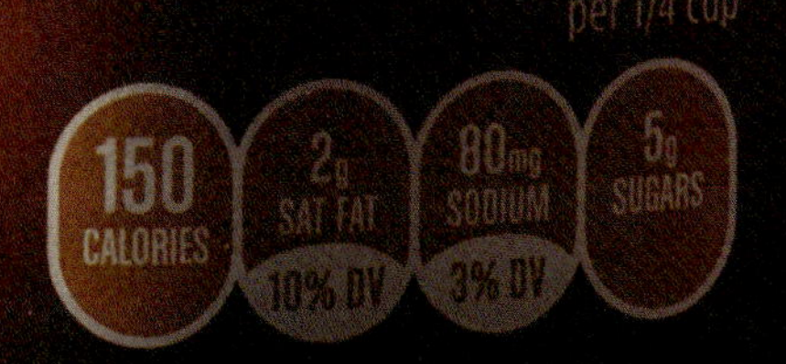
gold emblem
honey roasted
cashews
NET WT
8.25 OZ (234 g)
per 1/4 cup
150 CALORIES
2g SAT FAT 10% DV
80mg SODIUM 3% DV
5g SUGARS

PUBLIX
GreenWise®

ORGANIC PRESERVES
STRAWBERRY

Glaze your toast with an overly generous ruby-red slathering of the sweetest summer memories.

NET WT 12 OZ (340g)

PUBLIX

GREENWISE

★ 18 ★

In late 2000, Lakeland, Florida-based grocer introduced a new Private Brand – Publix GreenWise Market. The brand is built on the core belief that shoppers are looking for options that are more natural and less processed. They're trying to avoid additives and chemicals, and seeking products produced in a way they can feel good about. The idea was simple: create a brand customers will love that provides naturally delicious and environmentally friendly products. In order to be sold under the GreenWise brand, products must meet these standards:

"So let's all raise a bowl. Then tilt this box. Then well, you see where this is going."

- Organic items are raised without added growth hormones or synthetic antibiotics, steroids, pesticides, or fertilizers and are all natural with no GMOs (genetically modified organisms).
- All-natural items are minimally processed and have no artificial colors, flavor, preservatives or sweeteners.
- Earth-friendly products are created in a way that minimizes negative impacts on the environment.

In September 2007, Publix dramatically extended the brand and opened the first Publix GreenWise Market in Palm Beach Gardens, Florida. The 39,000 square-foot store offered customers a wide variety of health, earth-friendly, all natural, and organic products combined with high-volume traditional grocery items.

The latest manifestation of the Private Brand has dropped "Markets" from its name, opting instead for the simpler Publix GreenWise. At the same time, the decade-old package design has been re-thought. The new design is moving away from the clichés of grocery organic private labels; the retro inspired woodcuts and dominant green have been replaced by a fresh contemporary design that creates a visual and verbal voice that is playful, engaging and ownable. It is epitomized by the copy on the front of the Toasted Oats pack:

"We'd like to propose a toast to toasted oats. They're so crunchy and delicious, each spoonful turning your mouth into a bite-sized utopia of yummy goodness. So let's all raise a bowl. Then tilt this box. Then well, you see where this is going."

Publix GreenWise is a brand that confidently stands on its own two feet and proclaims its relevance both as a Private Brand and a retail concept.

GreenWise
ORGANIC SALSA
MILD
CERTIFIED ORGANIC
REJECT IF BUTTON IS UP
REFRIGERATE AFTER OPENING
CERTIFIED ORGANIC

PUBLIX
GreenWise®

ORGANIC CEREAL
HONEY NUT NATURALLY FLAVORED
TOASTED OATS

There's a nut in all of us. And there are many nuts in here. Not in an overpowering way. In a just right, perfect in your ice-cold milk kind of way. Tasty organic almonds and toasted oats all harmonizing with honey-flavored sweetness are just crazy good.

NET WT 14 OZ (397g)

HELP to OTHERS
H2O
PURIFIED DRINKING WATER
16.9 FL OZ - .5 LTR
100% RPET

HELP to OTHERS
H2O
PURIFIED DRINKING WATER
16.9 FL OZ - .5 LTR

H2O HELP TO OTHERS

★ 19 ★

In the spring of 2012, United Oil, a family-owned company that operates approximately 125 gas stations and convenience stores in locations throughout Southern California, introduced its very first Private Brand. The highly focused brand, H2O Help to Others bottled water, creates a daily opportunity for United customers to engage with the brand and help the retailer make a difference. With each bottle purchased, the retailer is donating 10¢ to local food banks to help people in the communities it serves.

> With great brand positioning and design, retailers can create brands that make a difference.

The Brand's up-to-date and engaging design brings to life United Oil's annual "I Got It!" Makes Cents fundraising event. One weekend a year, United Oil donates money, usually 1¢ to 2¢ from every gallon of gas sold at all of its gas stations, to a chosen charity. In 2012, United Oil wanted to keep the giving momentum going and came up with the concept to launch a Private Brand bottled water that would generate donation dollars every day of the year.

United Oil has demonstrated that with great brand positioning and design, retailers can leverage commodity products to create brands that make a difference.

HELPto
OTHERS™
H2O
PURIFIED
DRINKING
WATER
16.9 FL OZ - .5 LTR

HELPto
OTHERS™
H2O
PURIFIED
DRINKING
WATER
16.9 FL OZ - .5 LTR

HDX
Shower, Tub & Tile Cleaner
Removes hard water deposits and soap scum
Leaves no residue after rinsing
KEEP OUT OF REACH OF CHILDREN AND PETS.
DANGER: CAUSES EYE & SKIN BURNS.
HARMFUL IF SWALLOWED OR ABSORBED THROUGH SKIN.
See back panel for additional precautionary statements.
32 fl oz
32 FL OZ (1 QT) 946 mL

★ 20 ★

Bernie Marcus and Arthur Blank, along with investment banker Ken Langone and merchandising guru Pat Farrah, founded the world's largest home improvement specialty retailer, The Home Depot, in 1978. Their vision of one-stop shopping for the do-it-yourselfer came to life when they opened the first two stores on June 22, 1979, in Atlanta, Georgia. The retailer revolutionized the home improvement industry by bringing know-how and tools to the customer at a great price. Since then, the retailer has built an extensive portfolio of traditional private labels including Workforce, Hampton Bay, Commercial Electric, Glacier Bay and Husky, not to mention a significant collection of unbranded/generic, often value-focused products.

In 2012, The Home Depot shook up their private label portfolio with the introduction of a new Private Brand. Unlike previous labels, HDX leveraged a close connection to the retailer's brand name and iconic orange brand color and created a relevant multi-category "trusted value" brand solution Home Depot shoppers could grow to know and love. The brand is proof that inexpensive does not have to mean ugly or embarrassing. HDX is now found in a wide variety of value-focused home improvement products, including cleaning supplies, hand tools, shelving, bungees, fans, tarps and tape.

To support the launch of the brand, Home Depot challenged its associates to take part in an online video contest, the HDX Factor. According to the HDX Factor website:

"You've heard how great the new line of HDX products are; now it's your turn to pitch by creating your own amazing commercial or infomercial! All you have to do is make a 60-second or less video that convinces a panel of judges that your featured HDX product or products are absolute must-haves. You may even want to tell associates what the 'X' in HDX represents."

"The Grand Prize winner plus two associate guests (or submitting team up to three associates) will receive travel and accommodations to a VIP behind-the-scenes experience of an HDTV broadcast production in Atlanta, AND the winner's store will receive ONE THOUSAND DOLLARS for the Fun Fund!"

The clean, uncluttered design architecture and bright orange HDX mark creates a plainspoken, shopable brand that is easy to spot in Home Depot's vast aisles. HDX has become the dependable and durable brand solution for Home Depot customers.

HDX™
25 ft 7,62 m

HDX™
Scouring Pads
Almohadillas limpiadoras
.30 in W x 4.1 in L x 5.9 in H
.76 cm AN x 10.4 cm LA x 14.9 cm AL
6 pack

CONTAINS GLYPHOSATE
HDX™
Weed &
Grass
Killer Ready-To-Use
KEEP OUT OF REACH
OF CHILDREN
CAUTION
KILLS THE ROOT

HDX™
HDX™

100 watt replacement
CFL
light bulb
HOME 360
100 watt replacement
using only 23 watts
CFL
light bulb
1 soft white bulb
23 WATT

erasing pads
2 disposable pads
2
disposable
pads
HOME 360
clean dirt & grime
from a variety of
household surfaces
erasing pads

HOME 360

★ 21 ★

In 2008, Delhaize-owned grocers Food Lion, Hannaford, Bloom, Bottom Dollar, Sweetbay Supermarket, Harveys and Reid's were in the process of actively rethinking their Private Brand strategy and optimizing their portfolio. The dramatically consolidated set of brand assets included the introduction of the new household products brand, HOME 360. Debuting in Food Lion stores the same year, it quickly gained traction, as well as a PLBuyer private label packaging award.

From the latest designs in kitchenware to innovative cleaning supplies, the brand's black-based, contemporary design is accentuated by a clean, sans serif font, pops of primary colors and a design system that allows the brand mark to shift colors to fit the category and simplify the customer's shopping experience. The design marries stylish form with the promise of new and innovative products and creates a compelling brand experience.

> The design marries stylish form with the promise of new and innovative products.

Home 360 now covers a wide variety of products, including baby needs, pet care, and household products, with cool, modern designs and innovative products. The name is at once descriptive and emotionally evocative, promising that it covers the complete circle of customer needs, from diapers, wipes and baby lotions to pet food and kitty litter to household items like charcoal, cleaning supplies, stationery, paper goods, kitchen gadgets and seasonal items.

In 2010, the brand expanded beyond Food Lion to its Delhaize-owned bannermates, Maine-based Hannaford and Florida-based Sweetbay Supermarket and is now carried at all Delhaize-owned banners in the U.S.

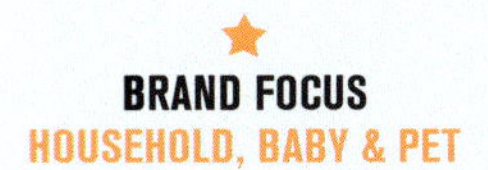

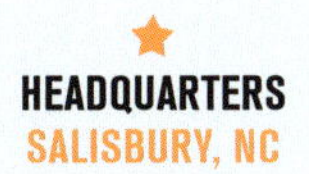

HOME 360™

HOME 360™
spring fresh
fabric softener sheets
40 Count - 6.4 IN X 9 IN (16 cm x 22 cm)
HOME 360™
to open - lift tab
spring fresh

143 CALORIES
3g SAT FAT
10% DV
173mg SODIUM
8% DV
9g SUGARS
PER 8 COOKIES

HT Traders
discover • inspire • enjoy

SEE NUTRITION INFORMATION FOR SATURATED FAT CONTENT
140 CALORIES
3.5g SAT FAT
20% DV
40mg SODIUM
2% DV
13g SUGARS
PER 2 COOKIES
Sir
CHOCOLOT
DARK CHOCOLATE TOPPED
TRUFFLE COOKIES
RASPBERRY
FILLED
NET WT 9.5 OZ (269 g)

HT TRADERS

★ 22 ★

In 2000, Charlotte, North Carolina-based grocer Harris Teeter launched their premium-tier Private Brand, H.T. Traders, with 62 items from Italy and Greece. At the time, the brand's tagline was "searching near and far" and focused on introducing the Harris Teeter customer to European and regional specialty products. Twelve years later, the brand has grown to hundreds of products and been repositioned to emphasize "fun, exciting, adventurous, and delicious." With the new positioning came a new logo, package design and corresponding fun and exciting brand voice, as well as the tagline, "Discover. Inspire. Enjoy."

The new branding emphasizes the fun and unique aspect of the product rather than its origin with packaging that reflects that. The rigid design system that was anchored by a dated globe-based logo has been replaced by a playful typeface that plays well with the "apple, fish and bread" art that is borrowed from the Harris Teeter logo. Every aspect of the brand reinforces its playful uniqueness including its product names ("Sir Chocolot" Dark Chocolate Topped Truffle Cookies, "Prairie Harvest Crunch" Multi-Grain Crisps) and category specific design. An ironic twist on the clichéd Private Brand "compare to" statement is found on the H.T. Traders Monumental Salted XL Virginia Peanuts and includes the phrase "*compare to the Pyramids" with a picture of the monument and a "Monumental" peanut Photoshopped in for comparison.

The design marries stylish form with the promise of new and innovative products

As the perfect example of confidence in brand, the distribution clause proclaims "Proudly Distributed by Harris Teeter."

Harris Teeter was aquired by Cincinatti-based grocer Kroger in early 2014.

HT
Traders™
discover • inspire • enjoy
PREMIUM HAND COOKED
MONUMENTAL
SALTED XL*
virginia peanuts
* compare
to the Pyramids

HT Traders
discover • inspire • enjoy
HT Traders
discover • inspire • enjoy
sweet & tangy
Carolina Mustard
BARBECUE SAUCE
HT Traders
general tso's
sauce
NET WT 12 OZ (340 g)

ONE
STEP

HY-VEE

ONE STEP

★ 23 ★

West Des Moines, Iowa-based grocer Hy-Vee launched Hy-Vee One Step as a sub brand of their mainstream line in 2012. The brand separates itself from traditional grocery private labels by confidently stating its purpose and reinforcing it with a design that delivers on its positioning:

> Together we can make a difference
> We know we can't solve all the world's problems.
> But we can do something. And you can do something, too.
>
> All it takes is one step. If we take that step together, we can help people in our communities and around the world overcome some of the difficulties that affect their well-being.
>
> The One Step mission is simple: To offer our customers a selection of everyday products and donate a portion of those proceeds to relevant, worthy causes.
>
> One small step to make lives easier, healthier, happier...One garden at a time.
> One meal at a time.
> One tree at a time.
> One well at a time.
> Won't you take that step with us?

According to their website each of the four products in the brand has a purpose and a charity.

Proceeds from the sale of One Step Russet Potatoes (5 lb. bag) help fund community gardens, teaching those in need about health and nutrition through the process of planting, tending and harvesting their own fruits and vegetables.

The Amos Hiatt Community Garden in Des Moines, Iowa, is the first garden developed with One Step funds used to develop a prototype that others can follow or modify to meet the needs of their neighborhoods and communities.

Hunger is a serious problem. Millions of people lack a dependable, adequate supply of food. Proceeds from One Step Shredded Wheat, packaged in a 100% recycled cardboard box, are dedicated to help those struggling with food insecurities.

Hy-Vee partnered with Meals from the Heartland, a non-profit organization of volunteers who package, transport and deliver highly nutritious meals to the needy.

Proceeds from the sale of One Step Paper Towels, which are made of recycled materials, will be used to make the world greener and healthier. In addition to reforestation projects, Hy-Vee will work with community organizations to plant trees in neighborhoods, parks and other public places.

The lack of a reliable source of fresh water is a global crisis. Worldwide, it is estimated one in six people lack access to safe drinking water; two in six lack adequate

HyVee
ONE STEP
A PORTION
THE PROCEEDS W
gardens
FOR COMMUNITIES
TO EAT HEALTHIER.
IT'S ONE SMALL STEP
one garden
at a time.
HY-VEE

CUSTOM-SIZE-IT
HyVee
ONE STEP
A PORTION OF THE PROCEEDS WILL
help plant trees AND ENHANCE OUR ENVIRONMENT.
IT'S ONE SMALL STEP, one tree at a time.
HY-VEE
112 2-PLY SHEETS PER ROLL
3 ROLLS

sanitation. Water-related illnesses are the leading cause of human sickness, suffering and death.

Using proceeds from sales of One Step Water, Hy-Vee is partnering with Rotary International to dig wells to provide millions of people around the world with better access to clean water and improved sanitation.

Many retailers talk about differentiation and change while simply playing the "compare and save" game. With One Step, Hy-Vee has created a true emotional point of difference.

19
LED TV
720p
60 Hz

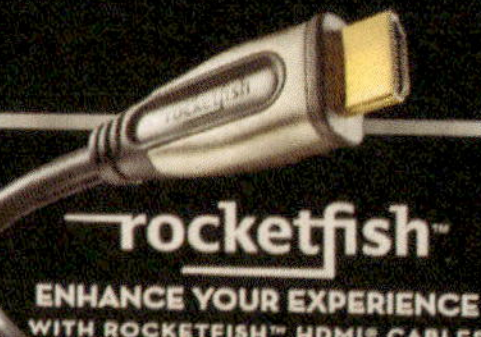

rocketfish™
ENHANCE YOUR EXPERIENCE
WITH ROCKETFISH™ HDMI® CABLES
(CABLES NOT INCLUDED)
AMÉLIOREZ VOTRE EXPÉRIENCE
AVEC LES CÂBLES HDMI DE ROCKETFISH
(CÂBLES NON INCLUS)
MEJORE SU EXPERIENCIA
CON LOS CABLES DE HDMI® DE ROCKETFISH™
(NO SE INCLUYEN LOS CABLES)
INSIGNIA™

INSIGNIA

★ 24 ★

In 2004, consumer electronics retailer Best Buy introduced its mainstream electronics Private Brand, Insignia, with a variety of television, DVD players, personal computers and accessories. Since then, Insignia has evolved from a line of inexpensive private label products into a fully formed consumer focused brand.

The Insignia website clearly and transparently calls out the "Uncommon Sense" of their brand promise:

> **The Insignia Difference:** To us it only makes sense that electronics should be dependable, affordable and do what they are supposed to. This notion seems like a no-brainer, but it is often lacking in the world of consumer electronics. That's why we call what we do Uncommon Sense.
>
> Being a part of the nation's largest electronics retailer gives us the insight necessary to know the perfect recipe for providing the highest quality products at an affordable price. A key ingredient in achieving this comes from listening to our customers, and gaining inspiration from those who spend the most time with our products.

Since its redesign in 2011, the brand voice and package design live up to the aspirational nature of its name — the minimalist design combines with a brilliant green and black color scheme and 3D logo that calls to mind sci-fi or gaming titles. The packaging exhibits a remarkably restrained information architecture highlighting the benefits and attributes the customer needs without cluttering the package with unnecessary information. The detailed, close-up product photography implies quality and an attention to detail. This combination creates an engaging, relevant brand that was recognized in 2012 with an American Graphic Design Award.

Over the last year, the brand has gained a newfound confidence and impressively extended both its products and marketing with new product extensions into HD radios, LED lightbulbs, the gigantic Insignia 65" Class LED 1080p HDTV, the second generation of the successful Insignia Flex Tablet, as well as a promotional agreement that highlights Insignia on TVs all across America as the official home theater partner of *The Ellen DeGeneres Show*.

POWERED
COMPUTER SPEAKERS
HAUT-PARLEURS
AMPLIFIÉS
POUR ORDINATEUR
ALTAVOCES
AMPLIFICADOS PARA
COMPUTADORA
INSIGNIA
NS-PCS20

8 feet
pi
pies
2,4 m
HD
1080p

INSIGNIA™
ENERGY STAR
LASTS OVER 22 YEARS
800 LUMENS
REPLACES 60W INCANDESCENT A19 BULB
LED LIGHT BULB

Kenmore

KENMORE

Kenmore, the now iconic Sears Private Brand, just celebrated its birthday 100 years after it made its debut in 1913, first appearing on sewing machines. In 1927, it appeared on an agitator-type, wringer washing machine and by the 1950s more than 10 million Kenmore branded products had been sold. Today, nearly one in three American homes contain a Kenmore appliance.

On March 12, 2010, the Kenmore brand launched its first interactive "Kenmore Live Studio" in Chicago. The studio is an interactive Private Brand experience that brings together live demonstrations, video and social media capabilities. Equipped with cameras that broadcast video via the Internet, studio visitors, chef demonstrations, and presentations and unveilings of new products are shared in real time with those following the Kenmore brand via its Facebook page. Unlike traditional Sears stores, the "Kenmore Live Studio" specifically promotes only Kenmore products.

In 2009, the brand underwent a total revamp, relaunching approximately 450 new or improved Kenmore appliances with an updated contemporary feel. Each new appliance was developed to fit the lifestyle of today's consumers with a focus on providing style and forward-thinking features and innovations that will change the way consumers interface with their appliances.

The redesigned branding included:

- A new Kenmore logo designed to symbolize the blend of the brand's rich heritage, a more modern, fresh image, and continued innovation leadership
- Updating the visual identity to include spirited language, bright colors, and simple yet current designs to help the brand stand out and demonstrate a bold, witty and fresh personality
- Leveraging the word "more" from within the brand name to create emotional interest: KenmorePop, KenmoreChill or KenmoreSizzle.
- New commercials with the tagline "Kenmore. That's Genius," which built upon the revitalized brand image while bringing the new brand vision to life in full color

Product design played a key role in Kenmore's reinvention. The brand created and incorporated a new product design philosophy that was carried throughout the development of all the new large and small appliance lines. The philosophy focused on streamlining state-of-the-art design to fit in any home, incorporating the new logo, custom font and sound palette while offering premium customer touch points and featuring user-friendly interfaces conveying premium technology.

The Kenmore brand, like Craftsman, continues to demonstrate that relevance, differentiation, loyalty and even love are possible in the best Private Brands.

BRAND
KENMORE

BRAND FOCUS
APPLIANCES

RETAILER
SEARS HOLDINGS

HEADQUARTERS
HOFFMAN ESTATES, IL

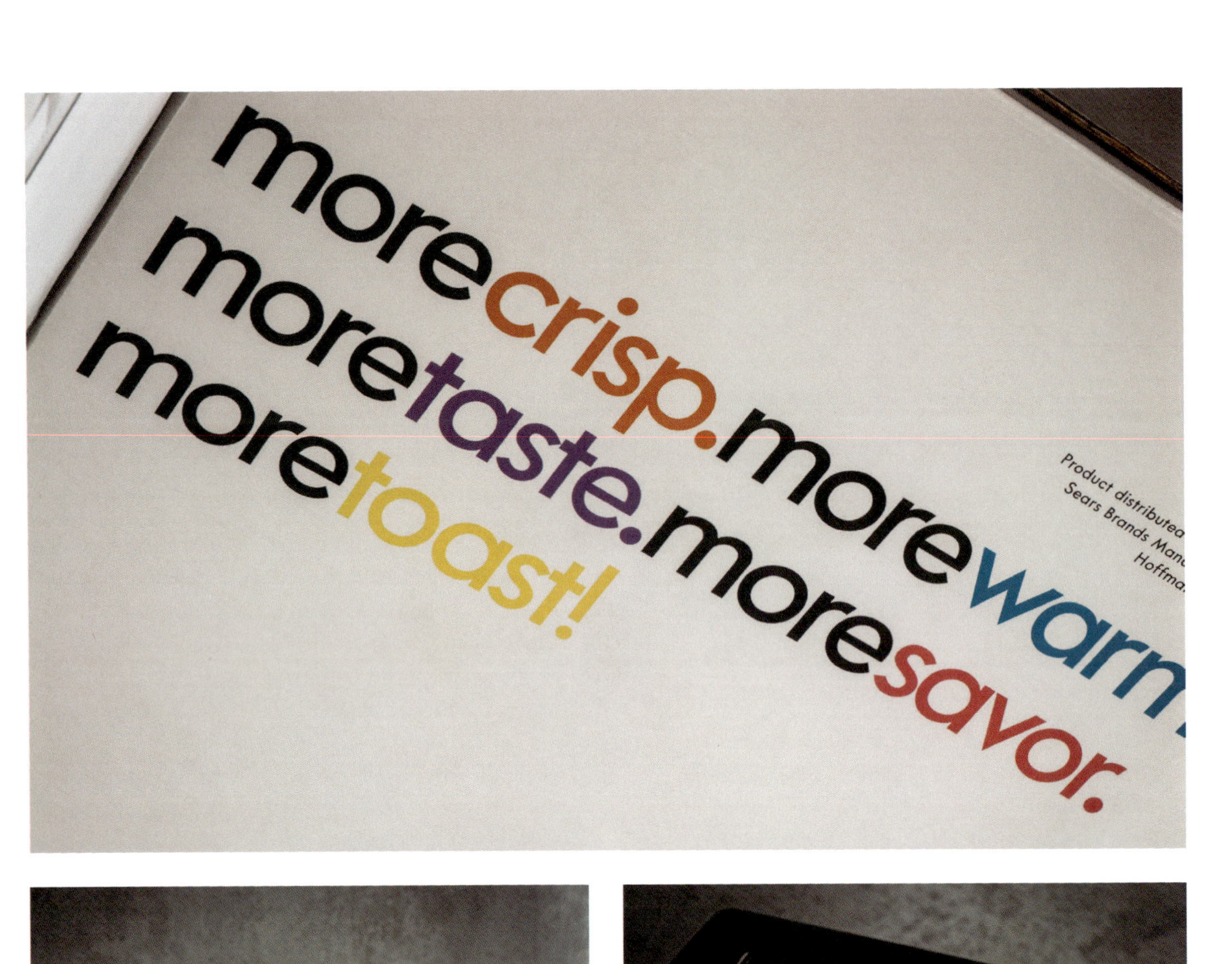
morecrisp.morewarm
moretaste.moresavor.
moretoast!

08 04203

Kenmoreperk!

Kenmore®

12-Cup Drip Coffeemaker

Cafetera de 12 Tazas

- Large, 12-cup carafe and quick-clean, non-stick warming plate
- Easy-to-read LED screen and push-button controls for a smooth start to your day
- Charcoal filter for a smooth, rich coffee taste
- Gold-tone basket extracts full, robust flavor

smooth.moreperk!

Kenmore
12-Cup Drip
Coffeemaker
Cafetera de 12 Tazas

Kidgets™
2 PACK
Receiving Blankets
26 IN x 26 IN
(66 CM x 66 CM)

KIDGETS

★ 26 ★

In June 2010, Matthews, North Carolina-based discount retailer Family Dollar strategically moved away from its traditional "me-too" private label strategy that was implemented in grocery and HBC (health, beauty & cosmetics) and created a new Private Brand. Kidgets, a $10-and-under clothing and accessories line for the newborn through toddler set, abandoned the category-focused private labels in favor of a brand targeted directly at moms. Designed to help her take care of her kids on a budget, the extensive collection of value-priced clothing and accessories encompasses virtually every category and product a busy mom might need, ranging from three-piece outfit sets and shoes to diapers and baby blankets.

The playful and emotive package design, naming and brand language abandon the generic and embarrassing to create a brand that moms can be proud to buy. Kidgets is not simply the functional basics that value shoppers can afford, but also includes the latest fashion trends at affordable prices. From complete outfits for infants and toddlers to must-have accessories including sandals, sneakers and jumbo diaper packs, Family Dollar has created a lifestyle brand in a sea of value.

To support the Kidgets launch, Family Dollar leveraged in-store marketing and signage, as well as a social media-based photo contest, "Cute as a Kidget." The contest awarded one Kidgets customer a grand prize of $5,000 in cash, a digital camera and accessories, and the opportunity to have their child featured in a Family Dollar advertising campaign. The Cute as a Kidget website also hosted a supplementary sweepstakes which conducted drawings for additional prizes and daily Family Dollar gift card giveaways.

Family Dollar continues to expand the line and support it with social media marketing through "mommy blogs." Mommies seem to love the brand, bestowing glowing reviews on the January 2013 introduction of baby wipes.

Kidgets is not only a brand that delivers on the Family Dollar retail brand positioning but also adds value to the Family Dollar Private Brand portfolio.

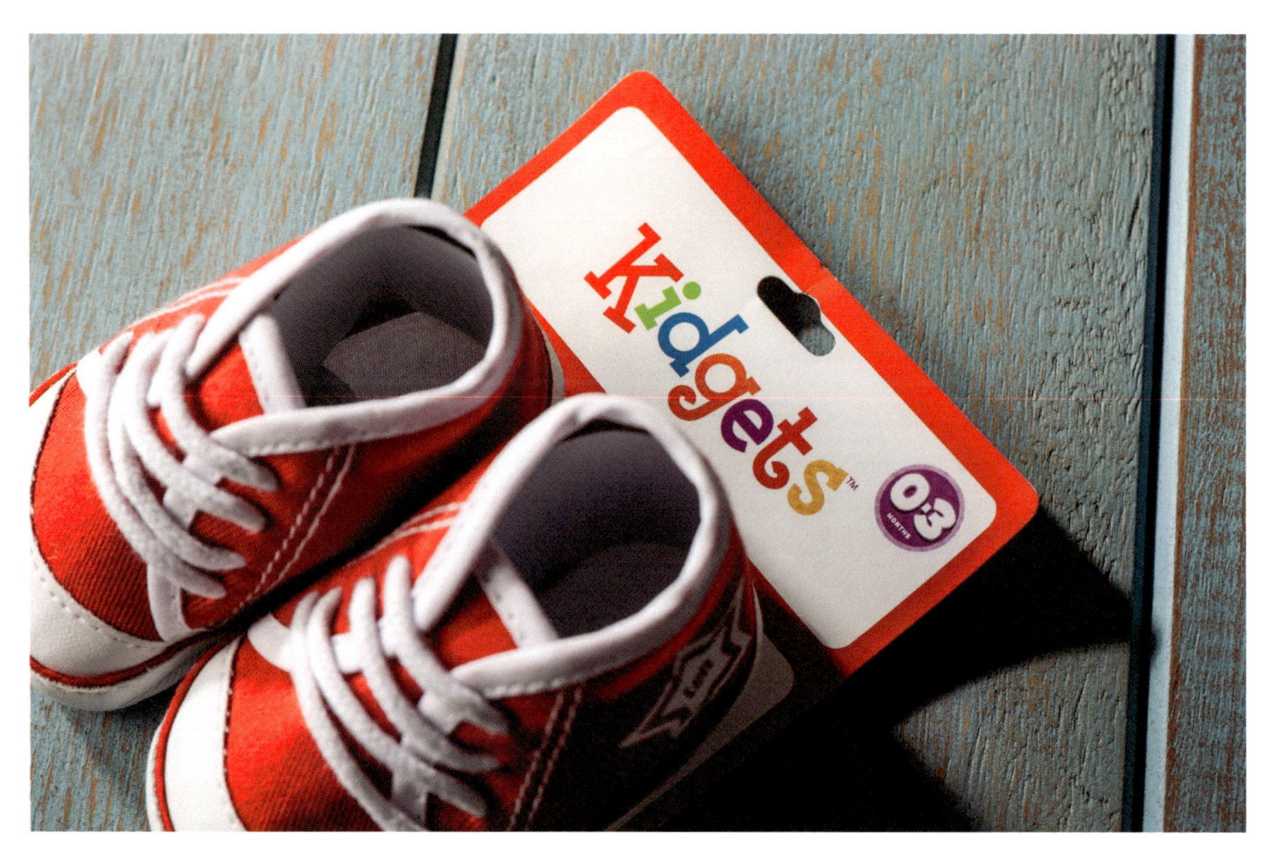
kidgets
0·3

Kidgets
6 PAIR
No-Show Socks
0·6
MONTHS
MADE IN CHINA

Kidgets™
FRAGRANCE FREE
Baby Wipes
Toallitas Para Bebé
PRESS HERE
SINGLE
DISPENSING
THICK • STRONG • SOFT
HYPOALLERGENIC • ALCOHOL-FREE
CONTAINS ALOE & VITAMIN E

KOBALT

KOBALT

★ 27 ★

Mooresville, North Carolina-based Lowe's Home Improvement launched Kobalt, the "Next Generation of Tough Tools," in 1998 as their Private Brand of mechanics' tools. Since then, the brand has extended to more than 1,700 products and become one of the fastest-growing tool brands in the United States.

In 2012, San Francisco-based branding agency Landor released its annual Breakaway Brands report. The study, published by *Forbes*, is designed to measure sustained growth in brand strength over a three-year period and includes a who's who of the world's greatest brands, including Facebook, Keurig, Skype, Amazon.com, YouTube, Netflix, and Apple. The 2012 list included three "Brands to Watch:" Kobalt, Foster Farms and Norton. The brands in this list exhibited significant upward momentum but did not make it into the top 10.

> Focused on engaging the suburban tool enthusiast, Kobalt and Lowe's are consistently winning.

Landor described Kobalt's emergence in this way:

"Sometimes a brand has to connect to consumers through one person. Kobalt, one of our brands to watch, has been able to involve its community with the strategic sponsorship of NASCAR driver Jimmie Johnson. As Johnson accumulated victories, and was voted Associated Press' Male Athlete of the Year in 2009, Kobalt gained increased visibility and positive associations. From sponsoring races to working with Johnson's team to design a line of tools, Kobalt's decisive partnership has allowed the brand to carve out a distinct identity."

By creating, managing and marketing an aspirational brand designed to professional standards and focused on engaging the suburban tool enthusiast, Kobalt and Lowe's are consistently winning. They have created a unique brand voice and visual language that is truly differentiated and ownable. The distinctive logo, branded hex pattern, ergonomic grip and Kobalt blue brand color have become a recognizable point of difference that Kobalt customers know and love.

#194170
KOBALT
LIFETIME HASSLE-FREE GUARANTEE™
3/16 in. x 4 in.
Slotted Screwdriver
Tournevis pour écrous à fente
Destornillador plano
Magnetic tip
Pointe magnétique
Puntas imantadas
KOBALT

KOBALT
#024007
Folding Lock-Back Utility Knife
Couteau universel à verrou repliable
Cuchillo para uso general plegable con bloqueo posterior
• Easy blade change mechanism
• Anti-slip aluminum handle
• Mécanisme de changement de lame rapide
• Poignée en aluminium antidérapante
• Mecanismo de cambio fácil de la hoja
• Mango de aluminio antideslizante
Easy blade-change mechanism
Mécanisme de changement de lame rapide
Mecanismo de cambio fácil de la hoja
5
Included/Incluses/Incluidas
KOBALT
KOBALT
Hooks to key chain
Peut être accroché à un porte-clés
Gancho para llavero
Cuts vinyl tile
Coupe les carreaux de vinyle
Corta baldosa de vinilo
LIFETIME

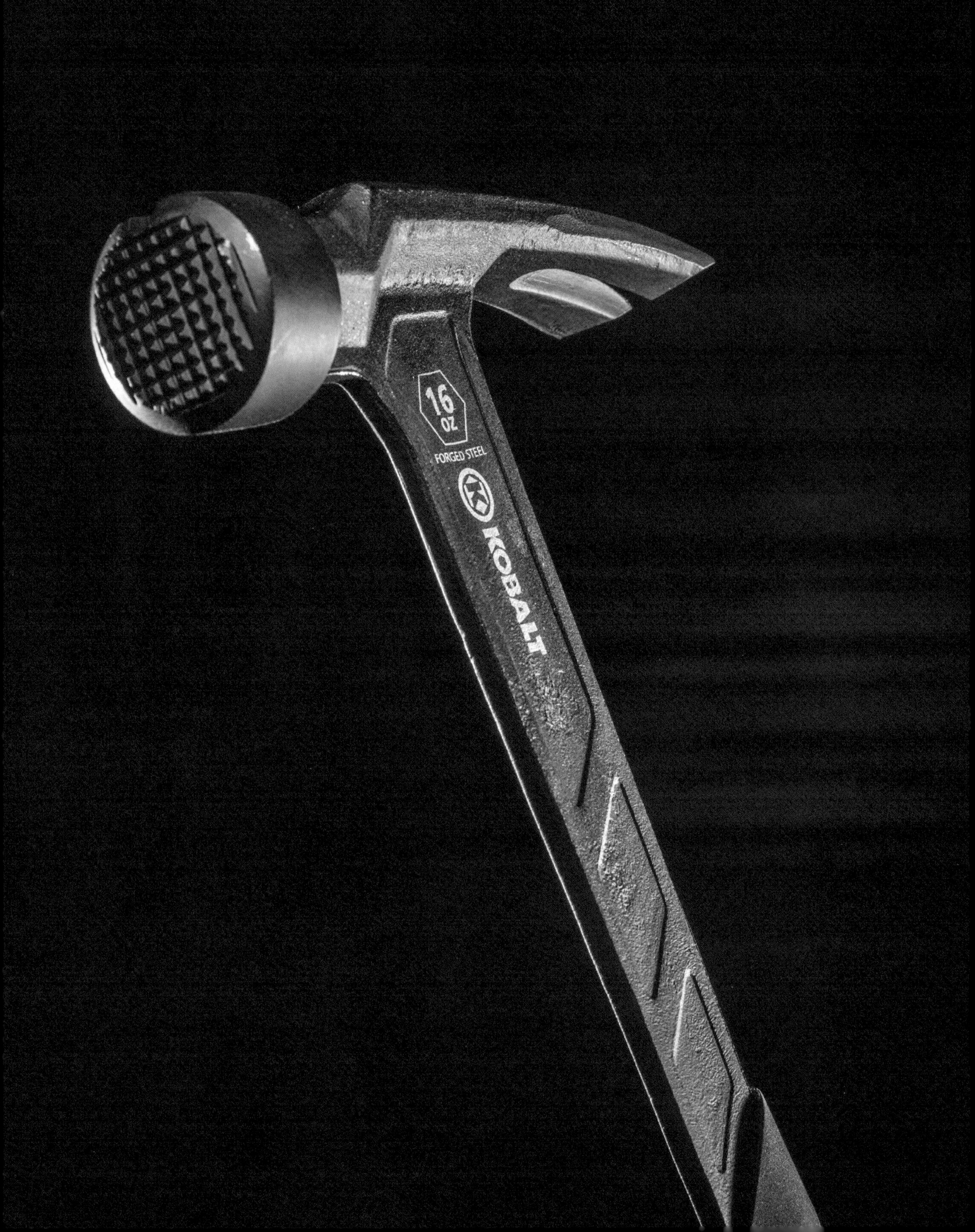
16
OZ
FORGED STEEL
KOBALT

LUCKY DUCK
LUCKY DUCK
chardonnay
south eastern
australia
FRANKLIN

LUCKY DUCK

★ 28 ★

In 2010, Bentonville, Arkansas-based retail giant Walmart was entrenched in the rollout of the ill-fated redesign of their largest grocery private label, Great Value. Unbeknownst to most industry watchers or customers, they also quietly rolled out a new Private Brand. Lucky Duck wisely targeted a younger demographic looking for a lifestyle experience rather than mere cost savings.

Lucky Duck was born with the playful name and design focused on conveying a fun, approachable personality that would appeal to millennial consumers who were usually intimidated by wine. They were looking for an easy to drink, affordable wine, which was also easy to shop. Lucky Duck provided a basic education and enabled the confident purchase of wine varietals.

The clean design highlights the wacky, irreverent duck and showcases the brand's international varietal range. Each expression and posture of the duck hints at the SKU's origin. Culturally specific ducks grace the Argentinian malbec, Chilean cabernet sauvignon and sauvignon blanc, and Australian chardonnay.

In 2010, the new Brand caught the eye of My Private Brand as well as the packaging design site *The Dieline*, and ultimately won awards from numerous design competitions including the 2010 *Print Magazine* Regional Design Annual and GD USA's 2010 American Graphic Design Awards.

Brands come in all shapes and sizes, and Lucky Duck is one brand that Walmart got right. With a Walmart-friendly price of less than $4, the wine's not bad either.

BRAND
LUCKY DUCK

BRAND FOCUS
WINE

RETAILER
WALMART

HEADQUARTERS
BENTONVILLE, AR

LUCKY DUCK
malbec
LUCKY DUCK
chardonnay
south eastern
australia

LUCKY
LUCKY DUCK
chardonnay
south eastern
australia
REGISTERED
FRANKLIN
U.S. PAT. OFF.

Marketside™

ALL NATURAL

CLASSIC CAESAR

DRESSING

NET WT 3 O

KEEP R

MADE FRESH DAILY!

Marketside™

CHICKEN

CAESAR SALAD

OUR CHEF'S RECIPE OF ROMAINE LETTUCE WITH TENDER

CHICKEN BREAST, BUTTER-FLAVORED CROUTONS, AN

PARMESAN CHEESE. INCLUDES CAESAR DRESSING

NET WT 14.5 OZ (411g)

MARKETSIDE

★ 29 ★

In late 2008, Bentonville, Arkansas-based retailer Walmart began Marketside, their grand experiment in small-format fresh foods grocery stores in the Phoenix metropolitan area. By 2011, the few remaining locations were shuttered, and the experiment came to a quiet close. However, the lasting and significant impact of the Marketside brand is still found in the Walmart of today.

In July 2009, the retailer began testing Marketside Private Branded products in its Walmart stores. Despite being overshadowed by the redesign and dramatic expansion of value positioned Great Value, Marketside has thrived and grown. The brand serves as the consistent, stalwart beacon of fresh quality in a retailer known for low prices but not great food. The Marketside brand and its package design speak to quality and freshness with a fresh, credible and approachable branded voice. It stands on its own as a brand and has successfully abandoned the copycat ugly design of Walmart's private label of old in favor of a unique and differentiated brand experience.

The brand now encompasses close to 100 products, including refrigerated meals, pizzas, breads and rolls, fresh pastas, salad dressings, bagged and prepared salads, and produce. Perhaps more interesting than the SKU count or categories are the unashamedly premium and seasonal products that bring the brand to life: Spinach artichoke hummus, fresh grilled chicken and spinach ravioli, garlic parmesan Italian Loaf bread, harvest blend salad, and fresh salsas, sauces and refrigerated salad dressings like sesame ginger, white balsamic yogurt and the seasonal pumpkin apple spice vinaigrette.

The Marketside brand demonstrates the potential for Walmart to not only create and manage great brands that give customers a reason to love the store for more than just low prices, but to leverage those brands to build credibility in categories and price points where they are challenged.

BRAND
MARKETSIDE

BRAND FOCUS
FRESH

RETAILER
WALMART

HEADQUARTERS
BENTONVILLE, AR

Marketside
SPICY GUACAMOLE
OUR AUTHENTIC GUACAMOLE WITH ALL NATURAL HASS AVOCADOS AND REAL JALAPEÑO KICK.
Hot
ALL NATURAL
2 SERVING TRAYS!
2 - 7 OZ Trays
NET WT 14 OZ (397g)
Marketside
CLASSIC SMOOTH GUACAMOLE
OUR ALL NATURAL AVOCADOS, TANGY LIME, AND GARLIC COMBINE FOR SMOOTH, DELICIOUS FLAVOR.
NATURALLY GLUTEN-FREE FOOD
Mild
ALL NATURAL
2 SERVING TRAYS!
2 - 7 OZ Trays
NET WT 14 OZ (397g)

ALL NATURAL
Marketside
BUTTERMILK RANCH DRESSING
OUR CLASSIC RANCH BLENDS SWEET BUTTERMILK AND SAVORY SPICES FOR A SMOOTH, CREAMY DRESSING.
ALL NATURAL

Marketside
FRENCH BEANS
ALL NATURAL
MICROWAVE IN BAG

Marketside
ALL NATURAL
CLASSIC CAESAR DRESSING
Marketside
MADE FRESH DAILY!
CHICKEN CAESAR SALAD
NET WT 14.5 OZ (411g)

Marketside™
BAKED FRESH DAILY!
FRENCH BREAD
OUR CLASSIC FRENCH BREAD
HAS A SOFT AIRY INTERIOR
AND CRISP GOLDEN CRUST.
FRESH BAKED IN-STORE DAILY.
NET WT 16 OZ (1 LB) 454g
Traditional

RENEW
mirra
CONDITIONING
LOTION
smooth and moisturize skin
cranberry & acai extracts
oleosomes
8 FL OZ (237 mL)
DAILY
mirra
MOISTURIZING
RENEW
mirra
AGE DEFYING
RETINOL NIGHT CREAM
0.83 FL OZ (24.8 mL)
RENEW
mirra
DAYTIME MOISTURIZER
BROAD SPECTRUM SPF 30

MIRRA

★ 30 ★

In early 2010, Cincinnati, Ohio-based grocer Kroger and its family of stores (Kroger, City Market, Dillons, JayC, Food 4 Less, Fred Meyer, Fry's, King Soopers, QFC, Ralphs and Smith's) leapt into the highly competitive beauty category with the introduction of the Private Brand, Mirra. The Brand wisely targeted "family-focused women seeking effortless beauty" and delivered with a name that evokes their morning ritual in front of the mirror. The packaging abandons traditional "me-too" private label design and structure in favor of creating an ownable and memorable Brand.

Mirra launched with 30-40 products and now includes a little more than 100. It combines natural ingredients with the latest science to meet three different consumer needs: Daily, for routine personal care; Renew, for a rejuvenating time out; and Inspire, for quick transformations for a night out.

For hair, there are Mirra Daily shampoos and conditioners available in three formulations: smoothing, volumizing and color treated. Mirra products contain sunflower extract, soy protein, amino acids and chicory root, along with other ingredients, to counteract the harsh effects of hard water and restore a healthy pH balance to hair. Styling products provide moisture, shine, and hold.

Mirra Renew skin care products feature an exfoliating cleanser and assorted creams to hydrate, reduce appearance of wrinkles, accelerate skin renewal, and firm and lift with ingredients such as aloe, green tea, pro-vitamin A and apricot seeds.

In body care, Mirra Renew offers bamboo & honey, lemongrass & ginger, black tea & cracked pepper and Japanese Yuzu fragrances in foaming bath, body wash, body lotion and body mist products.

RENEW
mirra
AGE DEFYING
RETINOL NIGHT CREAM
0.83 FL OZ (24.8 mL)
RENEW
mirra

RENEW
mirra
CONDITIONING
LOTION

RENEW
mirra
DAILY
mirra
MOISTURIZING
BODY WASH
refreshed skin
rosemary +
mint
quinoa protein
Vitamin A
18 FL OZ (532 mL)
CORRECTIVE SERUM

DAILY
mirra™
stand-up powder brush

GREEN TEA EXTRACT
COENZYME Q10
W
MYTRITION™
I AM
Beautiful. Brilliant. Balanced.
WOMEN'S
PERSONAL PACKS™
·30·
day supply
designed
specifically
for you
DIETARY SUPPLEMENT
GREEN TEA EXTRACT
COENZYME Q10
OMEGA-3 FISH OIL
M
MYTRITION™
I AM
Modern. Motivated. Masterful.
MEN'S
30

MYTRITION

The first Vitamin Shoppe store opened in 1977 on 57th Street in New York City. The new store served customers looking for healthy alternatives to over-the-counter and prescription drugs, and provided that with a variety of supplements, vitamins, and minerals. Less than 10 years later, in 1986, the first Vitamin Shoppe branded products appeared on shelf. In the more than 25 years since their introduction, more than 30 million Private Brand products have been sold.

In 2012, Vitamin Shoppe expanded their Private Brand portfolio with the introduction of Mytrition, a Brand designed to simplify the vitamin and supplement experience and conveniently fulfill unique lifestyle needs. Each Mytrition Personal Pack is designed to provide the benefits of the most advanced nutritional science in a convenient once-daily dose. Each Personal Pack includes a precise combination of high-quality tablets, capsules, and soft gels designed to provide customers with a wide spectrum of supplements, including vitamins and minerals, powerful antioxidants, targeted herbs, and essential fatty acids. Mytrition is available in Women's, Men's, Women's Sport, Men's Sport, Women's 50+, and Men's 50+ formulas, so rather than having to select, purchase, and organize multiple supplement products, consumers merely have to self-identify with one of the six segments, and all the work is done for them.

The brand has been so successful that in the Q1 2013 earnings call, Anthony N. Truesdale, CEO and director, announced that due to successful sales performance and positive customer feedback, Vitamin Shoppe would add four new SKUs by the end of 2014 which will grow the Brand by close to 35%.

The Mytrition name connects the Brand directly to a personal experience with the customer — while the aspirational language (Men's: Modern, Motivated, Mindful; Women's: Beautiful, Balanced, Brilliant) on each package reinforces an emotional connection with the Brand and a healthy lifestyle. The black and white customer segment relevant photography and fashionable yet classic logo and package design further reinforce the credibility and simplicity of the brand and create a true differentiator for the Vitamin Shoppe.

M
MYTRITION™
I AM
Modern. Motivated. Masterful.
designed specifically for you
MEN'S
PERSONAL PACKS™
·30· day supply
DIETARY SUPPLEMENT

W
50+
MYTRITION™
WOMEN'S 50+
PERSONAL PACKS™
M
SPORT
MYTRITION™

MULTIVITAMIN
100%
VITAMIN D3
M
SPORT
OMEGA-3 FISH OIL
L-GLUTAMINE
L-CARNITINE
GREEN TEA EXTRACT
L-ARGININE
MYTRITION™
MEN'S SPORT

W
50+
MYTRITION
Ageless. Audacious. Amazing.
WOMEN'S 50+
PERSONAL PACKS
30

W
MYTRITION
Beautiful. Brilliant. Balanced.
WOMEN'S
PERSONAL PACKS
30

M
SPORT
MYTRITION
Driven. Dynamic. Determined.
MEN'S SPORT
PERSONAL PACKS
30

M
MYTRITION
Modern. Motivated. Masterful.
MEN'S
PERSONAL PACKS
30

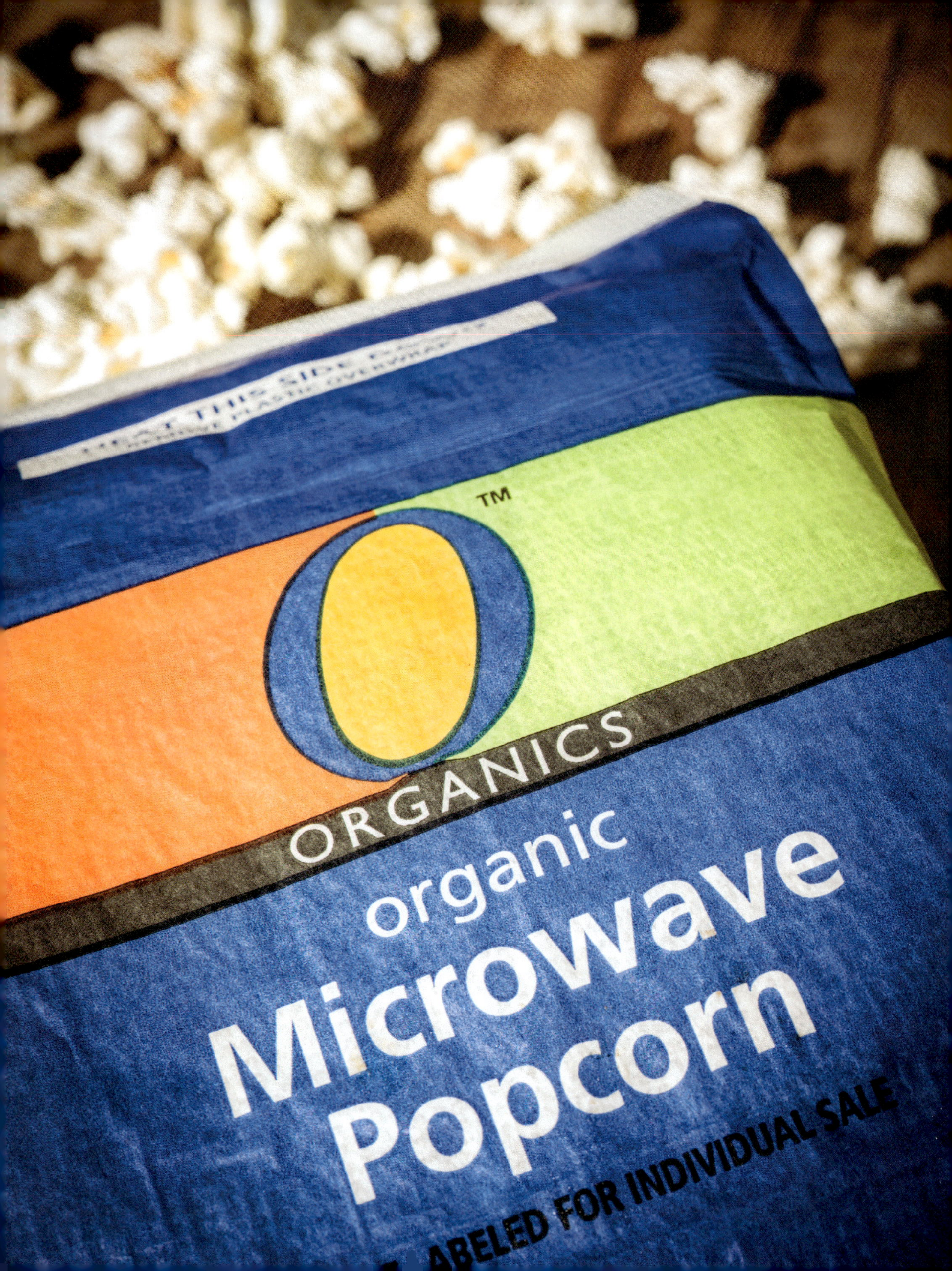

ORGANICS
organic
Microwave
Popcorn
ABELED FOR INDIVIDUAL SALE

O ORGANICS

★ 32 ★

In 2006, Pleasanton, California-based grocer Safeway introduced their now iconic organic Private Brand, O Organics. The brand quickly became the template for a whole generation of mainstream organic grocery Private Brands. Almost seven years later, the brand now has more than 400 items across virtually every category of the store, including baby & toddler; beverages; breakfast, dairy, eggs and cheese; frozen foods, fruits and vegetables; snacks and pantry staples.

The brand continues to reinforce its attribute-focused original brand message, "We believe that great tasting organic food should be available to everyone and sold everywhere at a great value."

O Organics' products are produced and handled in accordance with all USDA organic standards: without the use of synthetic pesticides, genetic modification, growth hormones or antibiotics. They are sourced from a variety of carefully selected organic growers using earth-friendly farming practices.

The Safeway website describes a few of the products this way:

> O Organics Fair Trade Certified Coffee supports a better life for farming families through fair prices, direct trade and community development.
>
> O Organics eggs come from free-roaming hens that are fed an organic diet that's grown without the use of pesticides or herbicides.
>
> O Organics produce is grown the way nature intended, without the use of synthetic pesticides and fertilizers.

Today, the brand name remains fresh and to the point. "O" now appears self-confident and fearless compared to the crop of generically named organic and natural brands that followed it, and the now-aging design walks the line between classic and dated.

O Organics has successfully carved out a place in the hearts — and on the dinner tables — of American grocery shoppers, and in doing so has become the standard by which highly focused niche Private Brands are judged.

BRAND
O ORGANICS

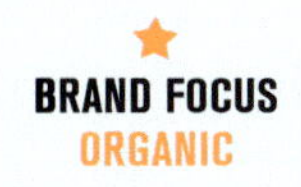

BRAND FOCUS
ORGANIC

RETAILER
SAFEWAY

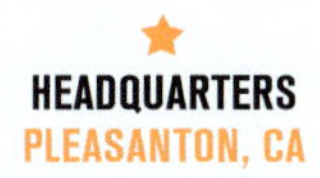

HEADQUARTERS
PLEASANTON, CA

O
ORGANICS
organic
oney Nut O
cereal

ORGANICS
Spearmint
O
ORGANICS
organic
Black Tea with Lemon
NET WT 16 FL OZ (473mL)

O
ORGANICS
PER 1 CUP POPPED
organic
Butter Flavor
Microwave Popcorn
NATURALLY FLAVORED
3-3.5 OZ (99g) BAGS
NET WT 10.5 OZ (298g)

ORGANICS
organic
Preserve

ology™
well. being.™

-NO-
HARMFUL
CHEMICALS
ALL-PURPOSE
CLEANER
CITRUS BASED
Cuts Through Grease & Grime
Our all-purpose cleaner is a natural spray for kitchen counters, stove tops & most household surfaces to clean & shine. Contains no harsh solvents, dyes or ammonia, so it is safe for you & your family.
MEYER LEMON & LAVENDER
26 FL OZ (769 mL)

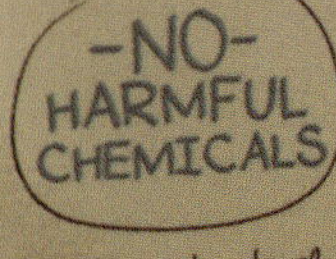
ology™
well. being.™
-NO-
HARMFUL
CHEMICALS
Soothing
BABY
SHAMPOO &
BODY WASH
ALL-IN-ONE
Gentle and natural formula to cleanse and protect baby's delicate skin & hair.
LAVENDER & CHAMOMILE

• hypoallergenic
• safe & gentle
• calming
13 FL OZ (384.5 mL)

OLOGY

33

In late 2012, the largest drug store chain in the U.S., Deerfield, Illinois-based Walgreens introduced the new Private Brand, Ology. The brand followed in the footsteps of Nice!, Delish and Well at Walgreens, which were introduced as part of the reinvention of the Walgreens portfolio of Private Brands, to create customer-focused Brands that become true points of differentiation for Walgreens.

Ology became the first nationally accessible Brand whose products are formulated to be free of harmful ingredients.

The new line of healthy household and personal care products combined the best of green, organic and natural to create a brand positioning that moves beyond the traditional constraints of merchandising categories and creates a customer-focused solution.

The text-based visual language and distinctive name reimagines the clichéd traditional language of "green" and targets the growing group of eco-conscious customers. The Brand is then brought to life on-shelf with unique engaging packaging structures and a tan and magenta color palette that begs to be touched, shopped and ultimately, purchased.

Ology, features a wide range of products, including paper towels, toilet paper and tissues made from bamboo and sugar cane husks instead of tree pulp. Other offerings in the initial 25-item launch included laundry detergent with fewer chemicals than regular brands, shampoos and conditioners for children and adults, compact fluorescent lightbulbs, and glass and all-purpose household cleaners.

BRAND
OLOGY

BRAND FOCUS
BETTER FOR YOU

RETAILER
WALGREENS

HEADQUARTERS
DEERFIELD, IL

ology
well. being.
NO HARMFUL CHEMICALS
PEPPERMINT
PURE CASTILE
SOAP
ALL PURPOSE
• daily body wash
• face wash
• shampoo
Everyday Multi-Use
32 FL OZ (946 mL)
ology
well. being.
NO HARMFUL CHEMICALS
MILK & HONEY
BODY LOTION
Naturally Moisturizing
• 24 hour moisture
• for normal to dry skin
8 OZ (227g)

ology
well. being.
NO HARMFUL CHEMICALS
Gently Soothing
FACIAL
TOWELETTES
HYPOALLERGENIC
• with aloe & green tea
FRAGRANCE FREE
85578-121

ology
well. being.
-NO-
HARMFUL
2X FABRIC

ology

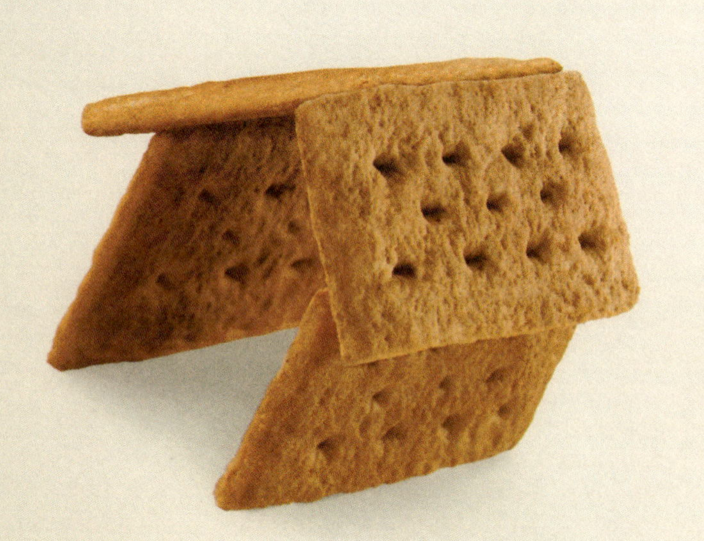
publix
HoneyGraham
SNACK CRACKERS
HONEY, I'M HOME.
PAREVE
NET WT
14.4 OZ (408g)

PUBLIX

★ 34 ★

Over the last few years, more retailers than I can count have rushed to follow Lakeland, Florida-based grocer Publix down the path of "white packaging" with most producing awkward or sometimes ugly copies. Publix first launched the now iconic Private Brand design in 2003. A clean, contemporary, predominately white monolithic design system, some ten years later it is still the best example of the genre. The groundbreaking design strategy not only threw away the traditional private label mentality of mimicking national brand designs, it also introduced the idea that a Private Brand should reinforce and build on the banner brand.

In addition to creating great designs, Publix truly uses their Private Brands as an extension of their brand. Each product reinforces and validates the retail brand in design, quality and value, all the while injecting a unique sense of humor. The designs are clean, elegant and well proportioned, utilizing white space to draw the consumer in, and creating clarity through their consistency.

The revolutionary brand strategy and design won Publix praise not simply from the typical trade publications like *Package Design* magazine and *Private Label Buyer*, but also from the prestigious graphic design business magazine *HOW*, which named Publix "in-house design group of the year" in 2005.

They continue to build and reinforce the brand with bold marketing and promotions that include in-store marketing, high-profile, high-penetration placement in their weekly circular, and the now famous "Store Brand Challenge" which gives shoppers who buy select national brands the opportunity to get the Publix brand version for free. They confidently proclaim, "We're so sure that you'll love our products, you can try them for free."

The Publix brand is evidence that a forward-thinking retailer can innovate through Private Brand and both engage the customer and lead the industry.

BRAND
PUBLIX

BRAND FOCUS
MAINSTREAM

RETAILER
PUBLIX

HEADQUARTERS
LAKELAND, FL

Publix
Black-Eyed
PEAS
NET WT 15.5 OZ (439g)
Publix
Cannellini
WHITE KIDNEY BEANS
NET WT 15.5 OZ (439g)
Publix
Red
BEANS
NET WT 15.5 OZ (439g)

OIL

Publix
Strawberry&Oats
SWEETENED MULTI-GRAIN CEREAL
Real Strawberries
10
NET WT 13 OZ (369g)

The Publix brand is evidence that a forward-thinking retailer can innovate through Private Brand and both engage the customer and lead the industry.

Calories
150
Per Serving
Total Fat
9g
Per Serving
Raley's
ALL NATURAL INGREDIENTS
ORIGIN
KETT
POTA
CHI

RALEY'S

★ 35 ★

When Raley's founder, Tom Raley, opened his first grocery store in Placerville, California in 1935, his dedication, hard work and vision quickly led the family-owned business to open stores in Sacramento, the great Central Valley, and eventually, Northern California and Nevada.

Raley embraced new ideas as quickly as he built new stores, introducing numerous innovations, including the first drive-in market, the first pre-packaged meat department, the first side-by-side grocery and drugstores (a "superstore") and the first supermarket natural foods department.

A good idea was a bright spark to Tom Raley's imagination, as long as it served his customers. "Treat our customers the way you'd like to be treated," he told his employees, "and they'll come back." That was — and still is — Raley's Golden Rule. That innovative philosophy inspired the introduction of the mainstream Raley's brand in the mid-1990s and its eventual redesign in early 2010.

The brand now encompasses more than 2,000 SKUs comprising such unique items as:

- Raley's Muesli, imported from Switzerland in five delicious flavors: Apple and Coconut Crunch, Honey and Wheat Crunch, Almond Crunch, Red Berry Crunch, and Chocolate Curl Crunch
- Locally crafted barbeque sauces made in small batches in the Sierra Foothills such as Smoky BBQ Sauce with Real Bacon, Raspberry Chipotle Barbecue Sauce and Spicy Roasted Garlic Barbecue Sauce
- Old world grains and rices including: Italian Farro, 5 Grains, Bulgur and Quinoa Mix, Barley, Peas and Lentils Mix and Basmati Rice, Oats and Lentils Mix

The brand includes a nod to Tom Raley and the continued ownership by the Raley family in the guarantee:

> As a family-owned company, our goal is to provide only the highest quality products that families can trust. In fact, we're so confident in this product — we stand behind it. If it doesn't meet your expectations please return it for a full refund. — The Raley's Family

The updated Raley's Private Brand design has dramatically transformed the brand from a traditional, if forgettable, private label to a unique and differentiated customer-centric brand. The warm color palette and multi-textured design maintain a loose unifying design strategy and information architecture while allowing each product and category to shine. Additionally, the front-of-pack nutritional icons provide a simple, easy to shop, transparent approach so that moms can make easy decisions at shelf for their families.

Raley's
swiss imported
müesli
apple and coconut crunch
made with whole grain
swiss quality
NET WT 12 OZ (340g)
Raley's
swiss imported
müesli
honey and wheat crunch
made with whole grain
swiss quality
NET WT 12 OZ (340g)
Raley's
swiss imported
müesli
almond crunch
made with whole grain
swiss quality
NET WT 12 OZ (340g)

Raley's
5 grains
mixture of spelt, barley, whole grain long rice, KAMUT® khorasan wheat and oats
made in Italy
NET WT 8.8 OZ (249 g)

Raley's
ALL NATURAL POTATO CHIPS
IT'S REAL BENEATH THE PEEL
WAVY WAVY

The Raley's brand has thrown off the transactional attribute focused shackles of "compare & save" private label and replaced them with a veracious curated Brand that works to actively and emotionally engage their customers. This is not an embarrassing private label; it is a self-assured brand moms can be happy to have on their shelves and tables.

Pull
To Open
the Snack Artist™
BEST BY
02 OCT 2013
YELLOW CORN
TORTILLA CHIPS
NET WT 13 OZ (369 g)
SEÑORITA CORNITA
PRODUCT ENLARGED TO SHOW TEXTURE

SAFEWAY

THE SNACK ARTIST

★ 36 ★

Over the last five years, Pleasanton, California-based supermarket Safeway has strategically and methodically reinvented both their Private Brand portfolio and the way they manage it. Chief marketing officer and executive vice president Diane Dietz spoke extensively about their Private Brand goals at the 2013 annual investors meeting:

"At the end of the day, it's about building loyalty… It's about creating unique brands that have our shoppers wanting to come to our store because they can't get the brand anywhere else. And so we are very focused on innovation."

One such innovation came in mid-2010 when Safeway introduced the sub brand The Snack Artist (exclusively for Safeway). The line included traditional potato chips, cheese curls, tortilla chips and crunchy kettle-style choices in a variety of cleverly named flavors like nacho cheese tortilla chips that brought the unique and playful brand voice to life with "Mariachicheese." Each of the snacks was named with a clever play on words and featured a fun, hand-drawn illustration that brought the "Snack Artist" to life on the package.

The brand has since expanded beyond salty snacks to encompass a wide variety of snack foods, including cakes, trail mixes, nuts, dips and frozen appetizers, each one evoking the fun the "Snack Artist" embodies.

The brand was well-received by both customers and critics alike and received numerous awards, including 2010 Pentawards, Silver Award: Food Packaging; the 2011 Communication Arts Design Annual 52, Award of Excellence: Package Design; and *Print Magazine's* 2012 Regional Design Award.

Less than three years later, the brand has quietly rolled out a redesign that simultaneously maintains the brand's spirit and amplifies its shelf presence.

Changes include:

- Bright yellow replaced the original tan background.
- "The Snack Artist" brand mark is dramatically larger in size and moved to a prominent location on the top left-hand side of the pack.
- The Safeway logo was de-emphasized and shifted to bottom left-hand side of the pack.
- The product illustration was reduced in size and now appears to be doodled on a white napkin and placed at an angle on the pack.

Overall, The Snack Artist is a confident, whimsical and fun brand in a category dominated by national brands and poor private label knock-offs.

the Snack Artist
An Oatie
But A Goodie
Exclusively for
SAFEWAY
OATMEAL
CRÈME
COOKIE
WITH CREAMY
FILLING
8-1.15 OZ (33g) COOKIES
NET WT 9.2 OZ (261g)
8 INDIVIDUALLY
WRAPPED COOKIES

the Snack Artist
PIGS IN A
BLANKET
Pork & Beef Hot Dog
Wrapped in a Light
Pastry Puff
NET WT 8 OZ (227 g)
PIGS IN A BLANKET
Pork & Beef Hot Dog Wrapped in a
Light Pastry Puff
BEST BEFORE
NOV 06 14
STN: SA463

the Snack Artist™
MORE SWEET & SALTY FLAVOR!
KETTLE CORN
MICROWAVE POPCORN
PRODUCT ENLARGED TO SHOW TEXTURE
EXCLUSIVELY FOR SAFEWAY
3–3.2 OZ (91 g) BAGS NET WT 9.6 OZ (272 g)

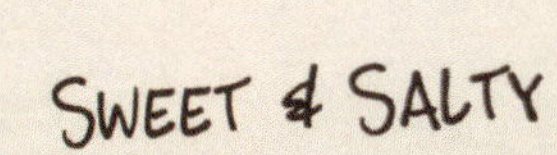
SWEET & SALTY

SEPHORA
Super lisseur
rides nuit
Age defy night
moisture cream
SEPHORA
yeux waterproof
Waterproof eye
makeup remover
50 ml 1.69 fl.oz.
SEPHORA
SEPHORA

SEPHORA

★ 37 ★

When Dominique Mandonnaud opened his first perfumery in 1969, perfume and cosmetics were hidden behind counters, jealously guarded by snooty sales people. With the opening of his first store in France, Mandonnaud revolutionized the way beauty was sold and created a new concept of retailing. Beauty products came out from behind the counters and became the star of the show. Perhaps most importantly, customers were encouraged to try, touch, smell, and discover.

> The eponymous brand has become Sephora's top-selling brand and now exceeds 1,400 items.

Today, beauty advisors offer advice and ideas that create a unique environment for experimentation and learning. Sephora's unique, open-sell environment features an ever-increasing amount of classic and emerging brands across a broad range of product categories including makeup, skincare, fragrance, hair care, bath and body products, as well as hair and makeup tools.

More recently, the retailer evolved from simply curating national and specialty brands to becoming a brand in their own right — one that's taken form in a powerful and compelling Private Brand. Sephora's Private Brand embodies all the brand's daring, creativity and excellence. It includes game-changing products that are as accessible as they are exciting, as sensual as they are superior. Like the 365 shades of lipstick — one for every mood.

The eponymous brand has become Sephora's top-selling Brand and now exceeds 1,400 items, which sell at the incredible pace of one every two minutes. The packaging design is clean, fashion forward and authoritative. Clean premium brand language is brought to life through the use of a sans serif font, as well as dominant black & white packaging complimented by color blocking and the occasional metallic highlight.

The Private Brand successfully extended to the Sephora stores inside JCPenney in October 2006 and is now included in 386 JCPenney stores nationwide.

BRAND
SEPHORA

BRAND FOCUS
BEAUTY

RETAILER
SEPHORA

HEADQUARTERS
SAN FRANCISCO, CA

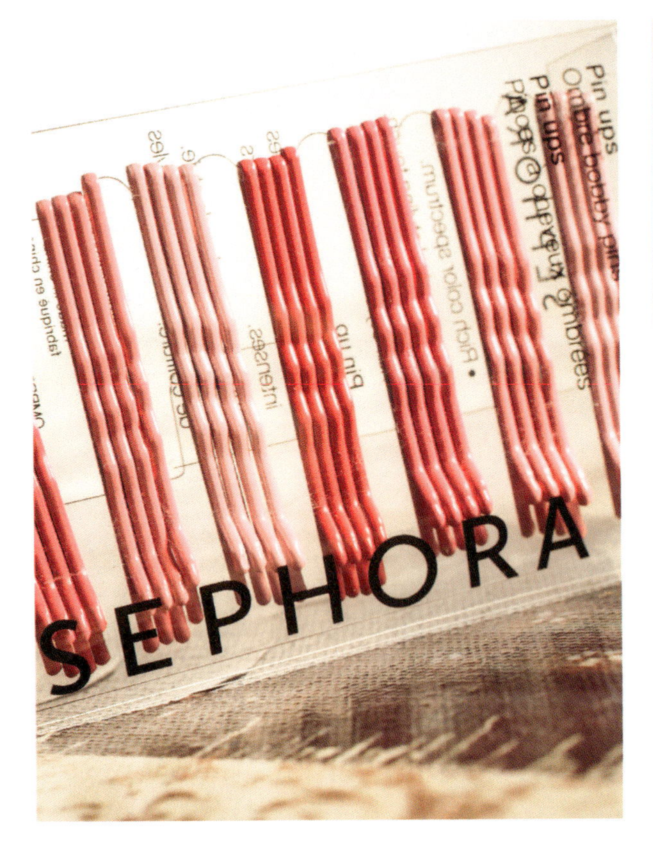
SEPHORA

SEPHORA

SEPHORA

SEPHORA
Super lisseur
rides nuit
j'en rêvais même la nuit !
Age defy night
moisture cream
sleeping beauty
50 ml 1.69 fl.oz.
SEPHORA
Super lisseur
rides SPF15
Age defy moisture cream
SPF 15 Sunscreen
don't act your age

FREE FROM
101
artificial
preservatives
& ingredients
CAFFEINE FREE
NEW!
TRY IT. YOU'LL LIKE IT!
simple truth
Blackberry
Pomegranate
SPARKLING BEVERAGE
ALL NATURAL FLAVORS
SERVING SUGGESTION
4 - 12 FL OZ BOTTLES (48 FL OZ)
4 - 355 mL BOTTLES (1.42 L)
150 CALORIES
0g TOTAL FAT
5mg SODIUM
36g SUGARS
PER SERVING

SIMPLE TRUTH

★ 38 ★

Launched in September 2012, Simple Truth is an intelligent strategic move by the nation's second largest grocer, Kroger, to optimize and consolidate their Private Brand portfolio, as well as create a Private Brand focused on customers' needs. The new brand eliminated the Naturally Preferred and Private Selection Organic brands in favor of the "Simple Truth."

Simplicity and truth inform every aspect of the brand from naming to product development to brand design and voice and tone. The positioning eschews the category functionality of traditional natural and organic Private Brands in favor of an emotional appeal to moms who simply want to take better care of their families. The traditionally complicated language of natural and organic is replaced by Simple Truth's 250 honest, easy and affordable "free from 101" items and certified-organic products designed for simply better living. All products are free from the 101 artificial preservatives and ingredients that customers told Kroger they didn't want in their products. Additionally, the USDA certifies Simple Truth Organic items organic.

The brand provides moms with a simple, uncomplicated solution to the challenge of better healthy living. Fresh, au courant packaging and easy-to-understand ingredient statements take the chore out of selecting Organic, Free From 101 and Natural foods. Spanning more than 30 product categories, Simple Truth and Simple Truth Organic products include a wide variety of foods, including milk, teas, salads, pizza, dried fruit, sodas, yogurt, chips and quinoa. Simple Truth Organic items display the USDA organic seal on the front of packaging, while Simple Truth products have highly visible identifiers that indicate their category.

Kroger and its family of stores (Kroger, City Market, Dillons, JayC, Food 4 Less, Fred Meyer, Fry's, King Soopers, QFC, Ralphs and Smith's) launched the brand with an unprecedented nationwide integrated marketing campaign, involving traditional media, in-store, and online components. Traditional media included television, radio and billboards. In-store communications include branded shelf signs, stanchions in produce and meat sections, and front-of-store standees and banners. Online elements included a Simple Truth website and a social media presence on branded Facebook, Twitter and Pinterest pages.

SELL BY MAR 16 13A
1%
simple truth organic
1% Low Fat Chocolate Milk
ULTRA-PASTEURIZED
HALF GALLON (1.89 L)
SELL BY APR 24 13A
2%
simple truth organic
2% Reduced Fat Milk
38% LESS FAT THAN WHOLE MILK
ULTRA-PASTEURIZED
HALF GALLON
Natural Cage Free, Grain Fed
LARGE WHITE EGGS | GRADE A
Unsalted Butter
GRADE AA
Unsalted Butter
GRADE AA

No artificial flavors or preservatives.
Black Beans
120 CALORIES | 0g TOTAL FAT | 125mg SODIUM | 1g SUGARS
PER SERVING
Tri-Bean Blend
130mg SODIUM
No artificial flavors or preservatives.
Diced Tomatoes
IN TOMATO JUICE
NO SALT ADDED
SERVING SUGGESTION
25 CALORIES | 0g TOTAL FAT | 15mg SODIUM | 3g SUGARS
PER SERVING
NET WT 14.5 OZ (411g)

BEST BY 26FEB2014
0099911 19:22
38g
whole grains
per serving
made with sliced apples, almonds & pecans
good source of fiber
contains 5g of total fat per serving
Simply balanced™
apple cinnamon muesli cereal
made with natural ingredients
Serving suggestion
NET WT 15 OZ (425g)

SIMPLY BALANCED

★ 39 ★

In mid-2012, Minneapolis, Minnesota-based retailer Target continued the evolution of its impressive and expansive Private Brand portfolio, with the spin-off of Simply Balanced from its parent Archer Farms. The launch included close to 250 products.

The new Simply Balanced collection replaced two subsets of the Archer Farms brand: Archer Farms Simply Balanced, which was launched in 2010, and Archer Farms Organic. The move quickly and impressively positioned the new Simply Balanced as a leading edge lifestyle-focused "better for you" and "free from" brand designed to take the guesswork out of eating well. Free of artificial flavors, colors and preservatives, the collection is built on purity, simplicity... and tastiness. It has become a true differentiator for Target and the branded solution that helps the consumer keep her life "Simply Balanced."

Simply Balanced committed to refrain from using 105 common food additive ingredients so consumers can be confident they're doing something good for their health and body. The vast majority of products within the collection are made without genetically modified organisms (GMOs) and as part of Target's commitment to wellness; the company is pledging to remove all GMOs from Simply Balanced by the end of 2014.

Simply Balanced sets itself apart in the Target Private Brand portfolio as the first grocery brand to step away from traditional private label tiering and fully embody the "design for all" ethos of Target — its colorful teal packaging and lifestyle-focused positioning are the embodiment of the Target brand in food. Simply Balanced is a credible and exciting combination of natural and organic, fashion and wellness that is uniquely Target.

Simply balanced
organic honey
all natural
USDA ORGANIC
NET WT 12 OZ (340g)

Simply balanced
organic trottole
enriched macaroni product
USDA ORGANIC

Simply balanced
5 count
Simply balanced
cherry almond Greek yogurt granola bars

Simply
balanced™
cherry almond
Greek yogurt
granola bar
OZ (35g)

CRUMBLED
SAUSAGE
PASTA SAUCE

traditional tomato sauce with mild Italian-style sausage

NET WT 25 OZ (1 LB 9 OZ) 708g

SIMPLY ENJOY

★ 40 ★

Ahold USA launched Simply Enjoy in their U.S. retail banners Stop & Shop, Giant Landover, Giant Carlisle, Martin's and Peapod in 2006. Since then, this premium-tier Private Brand has taken a different tone and path than the vast majority of the black-dominated, premium private labels from American grocers. Instead, they opted for a white background combined with rich photography.

However, since the beginning of 2013, the now seven-year-old brand has gradually introduced a radical redesign, setting a bright and fearless tone that clearly differentiates it from its competitors and sets a new standard for premium. The brand has created a positioning focusing on artisan-style foods, providing customers with unique and luxurious indulgences at great prices. The new brand design modernizes the Simply Enjoy logo with an updated sans serif font and a drop J that evokes the original logo. The tone-on-tone variation of the banner icon represented by a yellow bowl with three colorful halves that may be interpreted as bowls of fruit, bread, or ingredients subtly ties Simply Enjoy to the banner logos.

Behind every Simply Enjoy product is a story that brings the product to life. The citrus chamomile organic herbal tea promises:

"Refresh your day with our relaxing Citrus Chamomile Organic Herbal Tea. The infusion of peppermint, licorice and orange combine full-flavor with a soothing aroma in every sip. Start your day with a taste of Zen, or ease into your evening with a full-flavored palate cleanse. Any way you steep it you'll be sure to taste delight."

Simply Enjoy reinforces its brand promise with differentiated products and flavors including: Italian Crumbled sausage pasta sauce, raspberry-flavored shortbread cookies, rich chocolate lava cakes, creamy French Brie, citrus chamomile organic herbal tea, and spinach and artichoke tortilla crisp appetizers.

With an aspirational and evocative brand name, beautiful design and great product, Simply Enjoy proves that "Life is delicious."

BRAND
SIMPLY ENJOY

BRAND FOCUS
PREMIUM FOODS

RETAILER
AHOLD USA

HEADQUARTERS
WASHINGTON, DC

SIMPLY ENJOY
EARL GREY CRÉME
ORGANIC BLACK TEA
smooth citrus vanilla infusion
15 TEA BAGS
NET WT 1.32 OZ (37.5g)
RAINFOREST ALLIANCE CERTIFIED
OVER 50% CERTIFIED BLACK TEA

SIMPLY ENJOY
SPINACH & ARTICHOKE
TORTILLA CRISP
spinach and artichoke dip in tortilla cups

SIMPLY ENJOY
CHICKEN &
WHITE MEAT CHICKEN, BASIL

SIMPLY ENJOY
CHOCOLATE
LAVA CAKES
moist chocolate cakes filled
with rich chocolate ganache
SERVING SUGGESTION
FULLY

SIMPLY NOURISH™
DOG FOOD
CHICKEN & BEEF STEW
GRAIN & GLUTEN FREE
QUALITY NUTRITION
NATURAL INGREDIENTS
WITH ADDED ESSENTIAL VITAMINS & MINERALS

SIMPLY NOURISH

★ 41 ★

Since its launch in the spring of 2011, Simply Nourish from Phoenix, Arizona-based specialty pet retailer PetSmart has not only set the standard for Private Brand pet food, but has raised the bar for premium, natural pet foods across the board. Simply Nourish is a confident brand that mimics no one and takes a firm stand on product specifications and quality.

- Contains maximum levels of protein and minimum levels of carbohydrates
- Nutrient dense and easy to digest
- Contains no grains, soy or gluten
- Formulation that is easily digestible and nutritious

As a natural pet food, Simply Nourish is formulated with wholesome, real ingredients found in nature from high-quality sources with the nutrients still intact and free of artificial colors, flavors and preservatives. In addition, Simply Nourish meets Association of American Feed Control Officials (AAFCO) standards for natural pet foods. "Pet nutrition can be confusing and choosing the right food for your pet can be overwhelming," said Dr. Mark Finke, Ph.D. Nutritional Sciences and Nutrition Expert for PetSmart. "That's why Simply Nourish believes in making pet nutrition simple and understandable by using carefully-sourced, wholesome ingredients that are recognized by our pet parents, like real deboned chicken and a super food blend of real fruits and vegetables."

The brand name, wood cut iconography and color palette combine the familiar cues of natural and organic brands with contemporary type-based design. Together, they create a unique and ownable brand language designed to engage pet owners and create an ownable point of differentiation for PetSmart. The retailer has actively supported the brand since its launch with an impressive set of tools designed to build a branded experience.

Associate training played a big role and included a digital video to outline the brand story, brochure, sales contests and monthly newsletters with coupons for associates. Other tools included in-store marketing, a microsite, social media and an integrated campaign with the grooming department-which included a QR code driving shoppers to branded mobile touch points. The engaging "The Winning Ingredients" mobile video game was created to showcase the all-natural ingredients in every bag or can.

Simply Nourish natural foods are available for dogs and cats. Formulas for dogs include puppy, adult and limited ingredient options and are available in wet and dry options. Cat foods are available in wet and dry options in kitten, adult, limited ingredient, and indoor cat formulas.

S
N
SIMPLY
NOURISH™
Bisque Recipe
CAT FOOD - NOURRITURE POUR CHATS
CHICKEN & CARROT BISQUE
WITH PUMPKIN & QUAIL EGG
CAREFULLY SOURCED, QUALITY NUTRITION
MADE WITH
WHOLESOME
INGREDIENTS
SUCH AS REAL
DEBONED CHICKEN
NET WT 3 oz (85 g)

SIMPLY
NOURISH
Meatloaf Inspired Recipe
TURKEY RECIPE
WITH FETA CHEESE & SUN DRIED TOMATOES
NET WT 5.5 oz (156 g)

S
N
SIMPLY
NOURISH™
CAT TREATS
GÂTERIES POUR CHATS
GRILLED
CHICKEN

RESEAL FOR FRESHNESS

SMALL BREED
ADULT
DOG FOOD
NOURRITURE POUR CHIENS

CHICKEN &
BROWN RICE RECIPE

CAREFULLY SOURCED
COMPLETE NUTRITION WITH
OPTIMAL PROTEIN LEVELS
FOR AN ACTIVE LIFE

NATURAL
WHOLESOME INGREDIENTS
WITH ADDED ESSENTIAL
VITAMINS & MINERALS

REAL DEBONED CHICKEN
TO HELP SUPPORT A HEALTHY HEART & BODY

NET WT 6 lb (2.7 kg)

chocolate* covered
pretzels
best topping since
yellow mustard
CONTAINS
PURE VANILLA
9194 05535
2
*flavored

SKYLINE

★ 42 ★

The early months of 2008 were not good ones for New York City's hometown drugstore, Duane Reade. The retailer then owned by a private-equity firm, Oak Hill Capital, was struggling with debt, slow sales, dirty, cluttered stores and perhaps worst of all, unhappy customers. Despite having a huge presence in the New York area and a long history in the city, the retailer was unloved.

Duane Reade was going nowhere fast and desperately needed to reinvent itself to discover the true potential of the retail brand. Over the last five years, the story of the re-invention of Duane Reade and its subsequent acquisition by the drugstore giant Walgreens for $1.1 billion in February 2010 has become the epic story of the great American retail reinvention.

The new Duane Reade was laser-focused on bringing the retailer's new brand mantra of "New York Living Made Easy" to life with the optimization, consolidation and reinvention of the Private Brand portfolio. The Private Brand portfolio was to become a key strategic asset in their reinvention. The new portfolio: DR Delish in food, apt. 5 in cleaning and household products, the DR brand in health and beauty, and the un-named barcode or "skyline" brand each deliver on the New York sensibility.

The groundbreaking "skyline" brand features a consistent design that radically places the UPC code front and center. The bar code becomes artwork as it morphs from package to package taking on the shape of iconic New York City landmarks. The design solution is so simple. It elegantly supports the overall New York positioning of Duane Reade and creates an identity, a brand. A white-based brand with no name and no logo, it boldly proclaims its identity. This is proof that great strategy, vision, and design can make a difference.

Unfortunately, "skyline" was a victim of its own success, and after the acquisition of Duane Reade by Walgreens and the evolution of the Walgreens Private Brand portfolio, Walgreens' retail strategy and buying efficiencies meant the "skyline" brand would be phased out. As of this writing, there were only three SKUs left in the Duane Reade store on Wall Street and the will cetainly be gone by 2014.

spring
water
39194 05137
spicy
cajun
trail mix
hotter than
the subway
in summer
39194 05529
NET WT. 9 oz (225g)

mix
go ahead,
mingle with
the locals
OCT3111
KD7

starlight
mints
twinkle twinkle
little star
FAT FREE
6 39194 06840 6
NET WT. 8 oz (227g)

STONE RIDGE
Creamery®
SUNDAE NUT CONES
vanilla flavored ice cream with a peanut topped
chocolaty coating in a sugar cone
artificial flavor added
4 CONES
KEEP FROZEN
4-4.6 FL OZ (136mL) CONES / 18.4 FL OZ (544mL)

STONE RIDGE CREAMERY

★ 43 ★

Since the original trademark filing in 2005, Minneapolis, Minnesota-based SuperValu's Stone Ridge Creamery has been a solid, if expected private label. However, with a much-needed redesign in 2009, the private label emerged as an engaging and relevant brand. Stone Ridge Creamery combines a modern voice with one part hipster cool, one part small-town ice cream parlor and one part unique flavors to create a recipe for a Brand that customers love and that differentiates SuperValu.

With the tagline, "We love making ice cream," Stone Ridge Creamery proudly expresses a passion for food that moves it beyond old school "compare and save" private labels and creates an emotional, aspirational positioning.

> "We love making ice cream."

The Stone Ridge Creamery packaging goes on to further bring the brand story to life:

"A simple taste can take you back...

... Back to days filled with sunshine and bike rides to the local ice cream stand. Our quality ice cream, sherbets and yogurts can bring the same delight to your family just the way you remember it.

At Stone Ridge Creamery we love making and eating ice cream. It's our passion. We use only the best ingredients and don't take shortcuts. Whichever of our forty flavors you choose... there's a guaranteed smile in every scoop."

Over the last few years, Stone Ridge Creamery has created seasonal excitement through the introduction of limited edition seasonal ice cream flavors.

"Our summer limited-edition flavors — strawberry lemonade and harvest peach — were a tremendous hit with customers, and we're excited and confident that they'll have the same reaction to cinnamon snickerdoodle and peppermint cookies 'n cream and will want to include them in their holiday dessert menus," said Sam Mayberry, SuperValu's vice president of Private Brands.

SuperValu has impressively leveraged Stone Ridge Creamery as an important asset in their Private Brand portfolio, an asset whose value to the organization is defined not simply by margin or price but by differentiation.

STONE RIDGE
Creamery
REAL ICE CREAM
SALTED CARAMEL
1.5 QTS (1.42L)
KEEP FROZEN
ALL NATURAL FLAVORS
STONE RIDGE
Creamery
REAL ICE CREAM
Vanilla Ice Cream with Sea Salt Fudge and Chocolaty Truffles
SEA SALT FUDGE TRUFFLE
1.5 QTS (1.42L)
STONE RIDGE
Creamery
REAL ICE CREAM
Chocolate Ice Cream with Fudge Filled Chocolaty Fish, Caramel and Artificially Flavored Marshmallow Swirl
At Stone Ridge Creamery® we love making and eating ice cream. It's our passion. We use only the best ingredients and don't take any shortcuts. Each and every batch is slow-churned to creamy, delicious perfection. Maybe that's why our premium ice cream tastes so good. Whichever flavor you choose, there's a smile in
MARSHMALLOW GO FISH

STONE RIDGE

Creamery®

REAL ICE CREAM

Chocolate Ice Cream with
Fudge Filled Chocolaty
Fish, Caramel and
Artificially Flavored
Marshmallow Swirl

MARSHMALLOW GO FISH

1.5 QTS (1.42L)

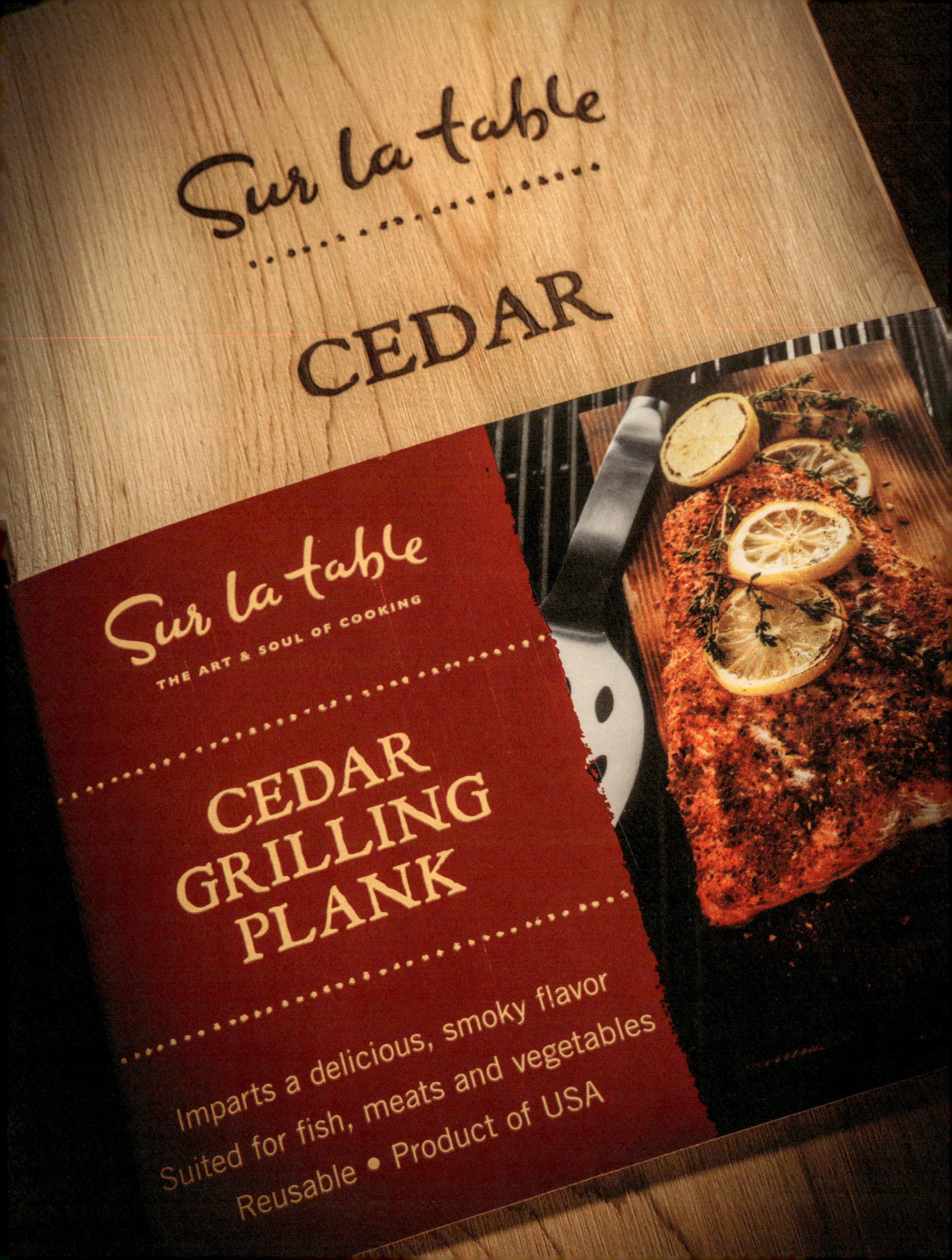
Sur la table
CEDAR
Sur la table
THE ART & SOUL OF COOKING
CEDAR GRILLING PLANK
Imparts a delicious, smoky flavor
Suited for fish, meats and vegetables
Reusable • Product of USA

SUR LA TABLE

★ 44 ★

Founded in 1972 by culinary aficionado Shirley Collins, Sur La Table was created in an era when it was difficult to find many of the specialty culinary tools Collins had found on her travels.

She began her dream with a small store located in Seattle's historic Pike Place Market, the oldest continuously operated farmers market in the United States. As a cook and food lover, Collins chose a location that put her store in the center of the market's fresh local produce and seafood.

Word quickly spread about the little shop packed to the ceiling with everything for cooks. Guided by Collins' culinary passion and expertise, Sur La Table's impressive selection of hard-to-find and cutting-edge utensils soon had food enthusiasts from around the globe making a pilgrimage to Seattle. Collins sold Sur La Table in 1995.

The retailer now offers the largest avocational culinary instruction program in the United States, teaching over 100,000 cooking enthusiasts a year how to cook. Skill levels range from beginner to advanced, with special courses devoted just to kids and teens.

Leveraging this heritage of cooking expertise, the retailer introduced the Sur La Table Private Brand with more than 2,000 Sur La Table branded SKUs on their website, as well as an extensive Private Brand assortment in the catalogue and stores. The brand has grown to encompass everything the cooking enthusiast needs to bring a great meal to the table, including cookware, cooking tools, linens, tabletop, grill tools and accessories, and imported and specialty foods.

The Sur La Table Private Brand retains the authenticity of Collins' original vision with contemporary and clean ingredient-focused design applied to products that cooking enthusiasts need and love.

Sur la table
HICKORY WOOD CHIPS
The simplest way to add wood smoke flavor to grilled foods
Kiln dried and all natural— no chemical processing
INS 28 CUBIC INCHES OF CHIPS IN A DISPOSABLE ALUMINUM TRAY • MADE IN USA
HICKORY WOOD CHIPS
IN A DISPOSABLE ALUMINUM TRAY
CALIFORNIA PROPOSITION 65 WARNING: COMBUSTION OF WOOD OR CHARCOAL PRODUCTS MAY CONTAIN CHEMICALS KNOWN TO THE STATE OF CALIFORNIA TO CAUSE
CANCER, BIRTH DEFECTS, AND / OR REPRODUCTIVE HARM. THIS WARNING IS REQUIRED AND ISSUED PURSUANT TO CALIFORNIA HEALTH & SAFETY CODE SECTION 25249

Sur la table

Sur la table

Sur la table
CAST IRON GRILL PRESS
surlatable.com

Sur la table
LENTILS
OF UMBRIA
NET WEIGHT: 17.6 OZ. (500 G)
Sur la table
THE ART & SOUL OF COOKING
FAGIOLI CANNELLINI
CANNELLINI
BEANS
PRODUCT OF ITALY
NET WEIGHT: 17.6 OZ. (500 G)
Sur la table
THE ART & SOUL OF COOKING
CECI ITALIANI
CHICKPEAS
PRODUCT OF ITALY
NET WEIGHT: 17.6 OZ. (500 G)

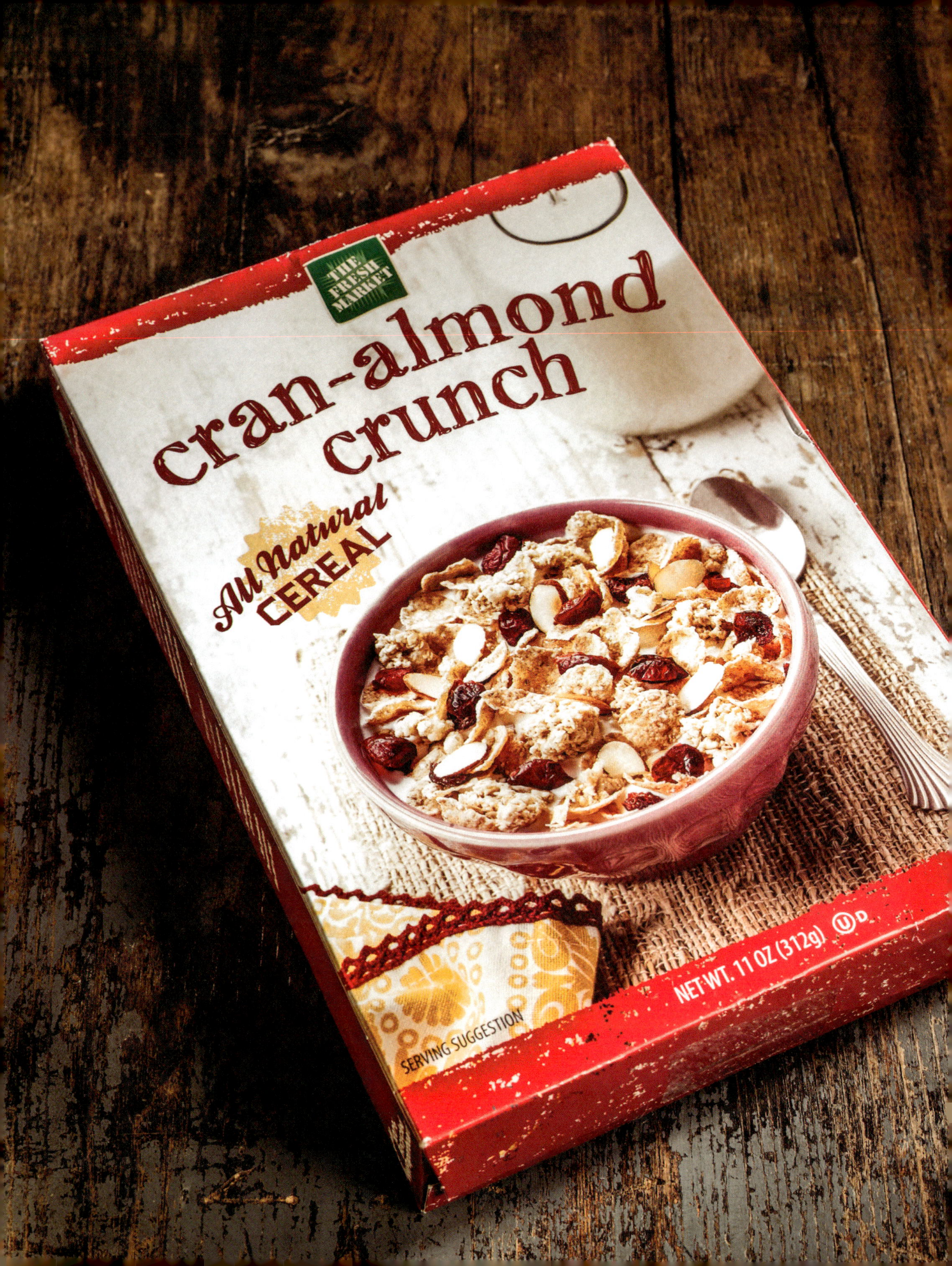
THE FRESH MARKET
cran-almond
crunch
All Natural
CEREAL
SERVING SUGGESTION
NET WT. 11 OZ (312g)

THE FRESH MARKET

★ 45 ★

On, March 5, 1982, Ray and Beverly Berry opened their passion — a new kind of grocery store — in Greensboro, North Carolina: The Fresh Market. The founders poured their life savings into their dream: a fresh take on the staid grocery formula. They created an experience designed to evoke the feel of open, European-style markets, celebrating fresh produce, meats cut in-store by a butcher, and unique and exciting packaged goods celebrating their ingredients.

Abandoning the over-commercialized, brightly lit, impersonal warehouse stores of the 1980s, The Fresh Market created a warm and inviting space filled with the best foods from around the world. Private Brand products quickly followed carrying The Fresh Market name and brand promise forward. Just as the stores abandon the sterility and sameness of traditional supermarkets, their Private Brand abandons bad design and the monotony of clichéd private label architecture and design in favor of impeccably executed, beautifully unique designs. Product photography brings the offerings to life and reinforces the old world sense of discovering a fabulous new product in the European market.

Since 2009, the brand has received close to a dozen design and product quality awards from some of the most prestigious competitions in the U.S., including *HOW Magazine* InHoWse Design Awards and the GDUSA American Package Competition, as well as Private Brand-centric competitions like the *PLBuyer* Packaging Awards and the Private Label Manufacturers Assocation (PLMA) Salute to Excellence Awards.

The Fresh Market has created a Brand that exceeds every traditional expectation of private labels and has become a trusted and tasty resource for cooking great meals and creating great food experiences.

SIGNATURE
SPICE RUB
ALL NATURAL
POULTRY
SPICE RUB
NET WT. 3.17 OZ (90g)

THE FreshMarket
WHOLE KERNEL
YELLOW CORN
THE FreshMarket
CUT
GREEN BEANS

THE FRESH MARKET
RIGHTEOUS
RED VELVET
CUPCAKE MIX
NET WT. 1.05 LB (16.86 OZ) 478g
serving suggestion

064 08 0614
THRESHOLD
QUALITY & DESIGN
12 PACK
SHOWER HOOKS
ONE SET

THRESHOLD

★ 46 ★

In mid-2012, Minneapolis, Minnesota-based big-box retailer Target undertook the biggest rebranding effort in the company's history: rebranding Target Home brand, the last Private Brand in the portfolio to carry the iconic Target Bullseye. In the transition, the retailer took the opportunity to revaluate every aspect of the Brand: positioning, name, logo, color palette, voice, product, quality, packaging, and style parameter. The new Brand, Threshold, debuted with an assortment of entertaining essentials, accents and decorative accessories and continued to expand through the entire home assortment in 2013.

The result is a bold and purposeful brand that expresses itself in an elegant, modern brand execution. The neutral gray and white accentuate the pop of green in the logo, as well as the details that make the packaging and design special. The subtle patterns and half-circle notch are thoughtful details that make for great brand design.

Beyond the package design, Threshold presents an intentional brand voice and design aesthetic that defines its sense of style in virtually every product. Color, texture, pattern and scale all combine with quality to create a brand that differentiates and gives the Target customer a brand and products she can love.

Threshold is the ultimate expression in home décor of Target's promise: "design for all." The brand is truly designed to express "her" lifestyle and create an asset in the Target Private Brand portfolio that differentiates with a strong personality, exceptional quality and beautiful product design. At their very best, Private Brands have the potential to move beyond the simple price, quality, and value equations and give customers a reason to choose one retailer over another, or one brand over another. Here, Threshold succeeds.

THRESHOLD
PULL SET
SOFT IRON FINISH
ONE SET
LABEL
LABEL

THRESHOLD
MINI BOWL SET
STONEWARE
DISHWASHER AND MICROWAVE SAFE.
THREE COUNT

THRESHOLD
ORGANIC BLANKET
100% ORGANIC COTTON FIBER
MACHINE WASHABLE

THRESHOLD
QUALITY & DESIGN
074 15 5091

TRADER JOE'S®
SPARKLING
natural LIME flavor
SPRING WATER
SODIUM FREE
REFRIGERATE AFTER OPENING

TRADER JOE'S

★ 47 ★

Trader Joe's is no ordinary supermarket; it's the rare American grocer that is also a fully articulated and well-defined brand. Their stores are offbeat lands of discovery that have reinvented the drudgery of grocery shopping and created a fully immersive brand experience.

The retailer's funky foodie roots and quirky in-store culture date to their founder, Joe Coulombe, who opened the first Trader Joe's in 1958 in Pasadena, California. The stores continue to bring their founder's vision to life. The retailer is heavily committed to its Private Brand, stocking shelves with a combination of value-focused organic and natural staples, as well as exotic and unique food discoveries. In short, it truly differentiates with its retail brands, its in-store experience, and its Private Brand.

Their Private Brand, unlike many of the private labels of suburban America's homogenous grocers, reinforces and builds on the uniqueness and brand personality of the Trader Joe's master brand. Through great product and unique brand packaging design, they have not only bucked the private label trend of monolithic white brand design, but have embraced compelling design to create something special. The Brand often appears as humorous sub brands focused on specialty foods or categories, including Trader José's, Trader Joe-San, Joe's Diner, Trader Ming's, Trader Giotto's, Trader Jacque's and Baker Josef's.

Their Private Brand design has eschewed the rigors of strict brand packaging design and architectural guidelines opting instead to create products that consistently bring their Brand personality to life. Although the retailer continues to embrace category-focused unique package design, over the last few years, the packaging has evolved to include a loose information architecture and color palette featuring consistent placement of the Trader Joe's logo, a decorative holding shape around the product name and a warm, complementary color palette. The Private Brand as a whole embraces a vintage, sometimes-kitschy feel that reinforces the master brand and encourages the Trader Joe's customer to enjoy the discovery.

BRAND
TRADER JOE'S

BRAND FOCUS
MAINSTREAM

RETAILER
TRADER JOE'S

HEADQUARTERS
MONROVIA, CA

TRADER JOE'S
ORGANIC
FAIR TRADE
BREAKFAST
BLEND
MEDIUM ROAST
MELLOW AND SMOOTH
100% ARABICA WHOLE BEAN COFFEE
NET WT. 14 OZ (397g)
TRADER JOE'S
JOE
... an exceptionally smooth cup of... coffee
MEDIUM ROAST
100% ARABICA WHOLE BEAN COFFEE
NET WT. 14 OZ (397g)
TRADER JOE'S
Organic
FAIR TRADE
Five Country
Espresso
blend
DARK ROAST
STRONG, RICH & WELL BALANCED
100% ARABICA WHOLE BEAN COFFEE
NET WT. 13 OZ (369g)

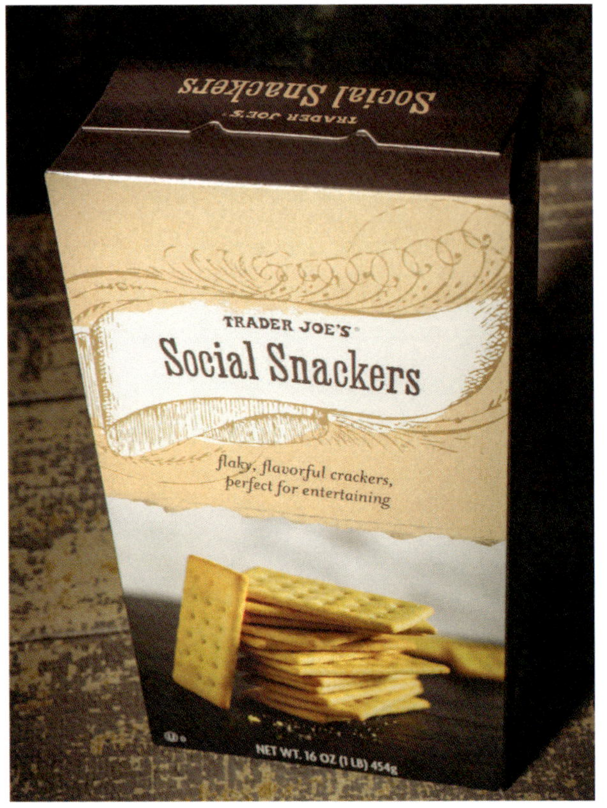
TRADER JOE'S
Social Snackers
flaky, flavorful crackers,
perfect for entertaining
NET WT. 16 OZ (1 LB) 454g

TRADER JOE'S
Social
Snackers
Trader Joe's Social Snackers
are so flaky and tasty that
you'll be tempted to eat them
right out of the box. If you
resist, you will have a great
foundation for a variety of
hors d'oeuvres. Top with
roast beef and Boursin
cheese with a sprinkle of
chives, or prosciutto with a
slice of Asiago and a small
piece of fresh fig-the
possibilities are endless. The
rectangular shape also makes

Organic

TRADER JOE'S®
Organic
Kidney Beans
Made with Sea Salt
USDA
ORGANIC

TŪL
ball
point
retractable
MEDIUM — BLACK
12
TŪL
TŪL
TŪL
TŪL
TŪL

In 2006, Naperville, Illinois-based office supply retailer OfficeMax launched the company's first design-based Private Brand, TUL. The groundbreaking brand included a selection of premium modernist design-inspired pens and dry-erase markers. With TUL, OfficeMax made the leap from selling "me-too" private label products to creating and managing brands.

The name, contemporary logo, design and visual voice are punctuated by the brushed metal brand color and sleek package design, which perfectly frames the product design. The retailer demonstrates remarkable restraint and focus, resisting the urge to clutter the package with the multiple features, attributes and callouts that plague traditional private label. With TUL's clear, minimalist packaging, the product shines and consequently, the brand is a star.

Six years after its introduction, TUL continues to grow and evolve. In May, OfficeMax announced plans to evolve two of its Private Brands, TUL and DiVOGA into national stand-alone brands, selling them at other retailers. In October, the retailer introduced the second-generation of TUL branded pens. The new line includes four new products that uphold the brand's reputation for delivering performance-driven and sophisticated designs.

"Over the past eight years, TUL has become the preferred premium brand of writing instruments and desk accessories for many of our customers because of the great design, performance and value it provides," said Ronald Lalla, executive vice president and chief merchandising officer at OfficeMax. "Our new collection of TUL pens provides both consumers and business customers with multiple designs that deliver exceptional writing experiences time and time again."

In 2013/14, OfficeMax merged with Office Depot.

TUL
fine point
permanent markers
12 - ASSORTED COLORS
7" straight scissor

TUL
highlighters
chisel tip — loop cap

TUL
ball point
retractable
MEDIUM — BLACK
12
ballpoint

TUL
dry erase markers

VIA ROMA
BOLOGNESE
AUTHENTIC ITALIAN PASTA SAUCE RECIPE
MADE WITH WHOLE PLUM TOMATOES
24 OZ
(680g)
PERSONALITY ★ ITALIAN FOOD
ITALIAN FOOD WITH PE
ANESCA
ROMA

VIA ROMA

★ 49 ★

In December 2008, The Great Atlantic & Pacific Tea Company (A&P), one of America's first supermarket chains debuted its groundbreaking Private Brand, Via Roma. The line debuted on filled-to-order cannoli in the bakery department, as well as a unique do-it-yourself cannoli kit, followed by the introduction of sauces, pasta, olive oil, cheeses, and pizza. It has grown to include more than 150 pastas, sauces, cheeses and other authentic Italian-style products.

The Italian foods brand brought to life the essence of the people of Tuscany — all different and all unique. Via Roma's uniquely Italian voice abandons the clichéd expression of Italian — red, white and green — in favor of emotive photography that shows the subject's true character, expression, and emotion.

Each package carries an emotive original black and white photograph commissioned by the grocer and portraying real and engaging older generation Italians, all taken in a small village in the Tuscany region of Italy. Combined with a stylish, upscale clean layout, typography and brand crest, the Brand embodies its tagline "Food with Personality."

> The essence of the people of Tuscany – all different and all unique

In 2010, the Brand was awarded more than a dozen design and branding awards from all around the world: 2010 REBRAND 100 Winners, The Communicator Awards, and *Private Label Magazine's* Store Brand Leadership Awards, as well as being featured on the influential sites *The Dieline* and *Lovely Package*.

The brand is now carried by A&P through its banners (A&P, SuperFresh, Food Basics, Waldbaum's, The Food Emporium and Pathmark), and continues to grow and prove the potential of Private Brands to emotionally engage.

ITALIAN FOOD WITH PERSONALITY
ITALIAN STYLE PRODUCTS OF THE HIGHEST QUALITY STANDARD AND PROCUREMENT
VIA ROMA™
BALSAMIC
VINEGAR
OF MODENA
ACETO BALSAMICO DI MODENA PGI
PROTECTED GEOGRAPHICAL INDICATION

VIA ROMA™
CHOCOLATE CHIP
BISCOTTI
AWARD WINNING & RESPECTED FAMILY RECIPE
NET WT 3.5 OZ (100g)

ITALIAN STYLE PRODUCTS OF THE HIGHEST QUALITY, STANDARD AND PROCUREMENT
VIA ROMA™
EXCLUSIVELY FROM THE TUSCAN REGIO

VIA ROMA
ANGEL HAIR PASTA
AUTHENTIC IMPORTED HIGH QUALITY ITALIAN ENRICHED DURUM WHEAT SEMOLINA PASTA
EXCLUSIVELY FROM THE TUSCAN REGION OF ITALY
"MY SISTER HAS HAIR LIKE SILK, BUT SHE WAS NO ANGEL GROWING UP. I THINK THAT'S WHY MAMMA WAS ALWAYS TRYING TO FEED HER MORE ANGEL HAIR PASTA, HOPING IT HAD AN EFFECT."
ITALIAN FOOD WITH PERSONALITY
VIA ROMA
ANGEL HAIR PASTA
AUTHENTIC IMPORTED HIGH QUALITY ITALIAN ENRICHED DURUM WHEAT SEMOLINA PASTA
16 OZ
TUCINE
16 OZ (1 LB) 454g

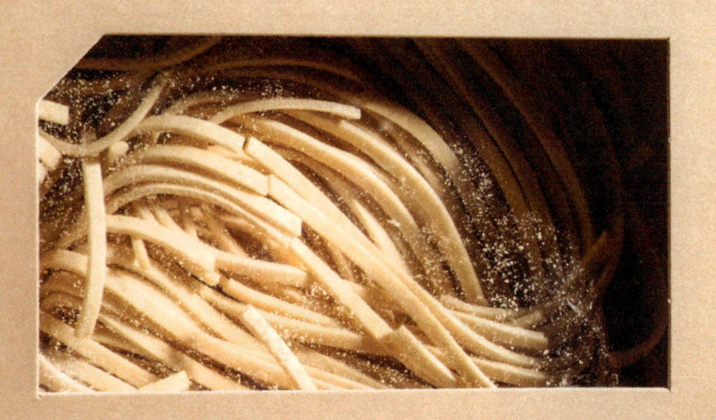
WILLIAMS-SONOMA
ARTISAN PASTA COLLECTION
SPAGHETTI
Hand-Cut
CRAFTED USING HEIRLOOM DURUM WHEAT
FROM WARD COUNTY, NORTH DAKOTA
NET WT. 1 LB. 1.6 OZ. (500g)

WILLIAMS-SONOMA

★ 50 ★

In 1947, Chuck Williams arrived in Sonoma, California, with the idea of building homes — first as a contractor, later as the owner of a hardware store, and finally as the visionary founder of the first Williams-Sonoma store for cooks. In 1953, Williams and a group of friends vacationed in Paris and discovered classic French cooking equipment like omelet pans and soufflé molds whose quality surpassed anything he had seen in the U.S.

According to Williams, "Cookware stores like Dehillerin or the housewares section of Bazar de l'Hotel de Ville in Paris were unlike anything most Americans had ever seen. I was completely fascinated not only by the vast array of kitchen tools and accessories, but also how they were displayed. Pots and pans in every conceivable shape and size, all out in the open."

In 1954, Williams purchased a hardware store in downtown Sonoma with the goal of converting it into a store, which specialized in French cookware. Within a few years, the building tools had been replaced by cooking tools, and the first Williams-Sonoma was born.

In the decades that followed, Williams-Sonoma expanded and became a cooking destination in towns and cities across America. During that same time, Williams-Sonoma Private Brand products debuted in stores and quickly brought to life the specialty, quality products that Williams discovered in Paris.

The simple, yet engaging package designs are current and fresh, each creating a unique brand statement that reflects the individual personality of the product. It builds on the overarching Williams-Sonoma brand by reinforcing the unique, artisanal, approachable and exceptional quality of the brand.

The Williams-Sonoma Private Brand serves as the ultimate expression of founder Williams' original vision. It has become the ultimate differentiator and the trusted Brand for both professional and home cooks to help them cook a great meal.

WILLIAMS-SONOMA
SHIMMERING SUGAR
NET WT. 4.8 OZ. (136g)

WILLIAMS-SONOMA
SAUSAGE
RAGU
A blend of San Marzano tomatoes, sweet Italian sausage, extra-virgin olive oil and basil
NET WT. 24 OZ. (1 LB. 8 OZ.) (680g)

CRAFTED MIXER
№01
Orange-Satsuma

WILLIAMS-SONOMA
NO. 1
ARTISAN PASTA COLLECTION
SPAGHETTI
Hand-Cut
CRAFTED USING HEIRLOOM DURUM WHEAT FROM WARD COUNTY, NORTH DAKOTA
NET WT. 1 LB. 1.6 OZ. (500g)

WILLIAMS-SONOMA
LIGHT, MOIST & TENDER
Coconut
For snowy white cupcakes with sweet shredded coconut
ALL-NATURAL
Cupcake Mix
CLASSIC RECIPE
ORGANIC FLOUR
MAKES 1 DOZEN
NET WT. 11.4 OZ. (324g)
LIAMS-SONOMA
Meyer Lemon
For rich, golden cupcakes bursting with sunny citrus flavor
Cupcake Mix
ASSIC
CIPE

Spiced Sweet
Potato Chips
Crisp and hearty sweet potato chips
seasoned with garlic and cilantro

WORLD TABLE

In March of 2009, the Private Brand industry was abuzz with the official launch of the redesigned and reformulated Great Value at Walmart. The renewed commitment to Great Value combined with SKU rationalization and a worsening economy spelled doom for the fatefully austere private label packaging, and the retailer was forced to backpedal and rethink its strategy.

At the height of the Great Value controversy, Walmart quietly introduced the premium brand, World Table. Silently entering the store, it replaced many Sam's Choice items and introduced new, unique and interesting products. And unlike its much-maligned sibling, Great Value, it abandoned the generic and established a distinctive brand positioning and visual voice with a brand story and package design that competently brought them to life. The packaging proudly reads:

"At World Table, we explore regional cuisines and international dishes, so you can bring these authentic tastes into your home every day. All of our recipes start with a variety of high quality ingredients and are crafted with care. You and your family will discover a world of new, delicious flavors in every bite."

Despite its foundation, the brand suffers from the lack of strategic importance created by the implosion of the Great Value private label strategy. Out of stocks and dropped products have become the rule rather than the exception.

SO WHY INCLUDE IT IN FIFTY2?
The answer is two-fold:

ONE: Walmart shoppers have proved that the brand lives in the heart of customers and not an office in Bentonville, AR. My Private Brand readers who are Walmart customers have submitted 100 comments consistently complimenting the brand and its products and complaining about the out of stocks.

"The offerings seem to be uniformly excellent with the Yukon Gold Garlic Mashed Potato chips topping the list!! The salsas, cookies, and snack mixes provide more healthful options with great flavor and variety. Go World Table!!" — Marsha

"Just tried the World Table Thin Crust Pepperoni Pizza with Fresh Mozzarella. Surprisingly good! The extra circles of mozzarella cheese are a definite upgrade and the pepperoni was excellent. So was the sauce. The end crust would actually fall apart in my mouth! A first for me from a frozen pizza." — Wally

"I love the World Table brand. The Spiced Sweet Potato Chips rock and I've loved all the crackers I've tried so

World Table
Bread Twists
A TASTE OF FRANCE
World Table
Bread Twists
Crunchy snack twists with a flaky rich, buttery flavor and a hint of salt
World Table
Waffle Crisps
Crispy, buttery cookies made from a classic Belgian recipe
A TASTE OF BELGIUM
NET WT 3.5 OZ (100 g)
Entertainment
Cracker
Collection
A premium assortment of classic European-style entertainment crackers
NET WT 8.8 OZ (250 g)
A WORLD OF FLAVOR
World Table
Entertainment Cracker Collection

World Table
Buffalo Style Chicken
Thin Crust Pizza
Seasoned white meat chicken, tangy buffalo-style sauce and real mozzarella cheese top this delectable thin-crust pizza
DELICIOUS
A TASTE OF USA
DISCOVERIES

far…Jalapeno Cheddar Crackers, Honey Flax Crackers, Black Pepper and Sea Salt Crostini Crackers. The biscotti are excellent, as well as the Spiced Chai and the thin sugar cookies. All super yummy and my food snob friends have changed their ways as well. I'm so happy to be able to get these great food items at my local Walmart." — Kimberly

"We were just at Walmart and tried to find the World Table Berry Crackers, but didn't see them in the cracker aisle at all. Really like them, but can't find them!" — Kay

"I love the World Table Cheese bites but can't find them much. We have been to 5 Walmarts in this area and only 1 had them. They are awesome!!" — Darleen

TWO: World Table, along with other brands in the Walmart Private Brand portfolio (Lucky Duck, Marketside, Ol Roy, George, Canopy, etc), demonstrates the potential for Walmart to create and manage great brands that give customers a reason to love Walmart beyond just price.

Walmart's World Table demonstrates that well constructed, competently designed, consumer focused brands have a place not only in the Walmart Private Brand portfolio but in the heart of the Walmart customer.

SELECT
Cranberry
Cocktail
130% daily value
per
16 FL OZ
SELECT
Orange
Cocktail
100% daily value
per

7 SELECT

★ 52 ★

In early 2011, Dallas, Texas-based 7-Eleven, the world's largest convenience store chain with more than 40,800 stores in 16 countries, redesigned and relaunched its primary Private Brand, 7-Select. Abandoning a 2008 traditional private label redesign, and despite the clichéd private label name "7-Select," the new design encapsulated the energy, vibrancy and on-the go nature of both the convenience channel and the 7-Eleven retail brand.

The Brand now includes more than 200 items with unique flavors that work to differentiate 7-Eleven. The website and packaging bring the Brand to life through a playful, irreverent and self-possessed brand tone. It is rare for a retail website to so effectively evoke a brand tone with its intended audience:

> **7-Select:** We want you to like us. A lot. So in the process of severely sucking up, we developed a product line that includes all your favorites. All high quality. All at a great price. Who needs national brands when you can be BFFs with 7-Select and have money left over to take us to dinner? Please?

> **LOGO ALERT:** When you see this logo on the website, stop whatever you're doing and check it out. It means it's a 7-Select product. And that means it's more important than anything else you could possibly be doing.

The lighthearted attitude continues on packaging with product descriptions that focus on flavor, quality, attitude and voice, and not the traditional "compare & save" message:

> **7-Select Big Bite Hot Dog Chips:** Make a note in your diary because your life is about to change forever. Introducing the Big Bite Hot Dog Flavored Potato Chip. Layer upon layer of hot dog flavor – mustard, ketchup, relish and hot dog – all loaded on each chip. Don't be surprised if you cry a little the first time you try them. They're that good.

> **7-Select Watermelon Sherbet:** Want to hear a juicy secret? This sweet, fruity sherbet is filled with tiny candy seeds. Fresh from 7-Select where the price is small, but the taste is mighty.

The 7-Select brand now stands toe-to-toe with the retailer's iconic Private Brands: Slurpee, Big Gulp and Big Bite hotdogs, and reinforces the 7-Eleven brand.

Cheddar & Sour Cream
FLAVORED POTATO CHIPS
SELECT
Sour Cream & Onion
FLAVORED POTATO CHIPS
Classic. Thin and crispy.
0g Trans Fat • No Cholesterol
BIG BITE Hot Dog FLAVORED
POTATO CHIPS
Classic. Thick and crispy.
0g Trans Fat • No Cholesterol
NET WT 2.5 OZ (70.8g) / CONT. NET. 70.8g

SELECT
Lemon
CREME COOKIES
With creme filling.
A sweet and tasty treat.
Best By
MAR 12 14
Duplex
CREME COOKIES
With creme filling.

Hot Chocolate
ICE CREAM
Spiced chocolate ice cream with marshmallow swirls
NATURAL AND ARTIFICIAL FLAVORS
ONE PINT (473mL)
Homemade Strawberry
ICE CREAM
Creamy homemade strawberry ice cream with sliced strawberries
ALL NATURAL
ONE PINT (473mL)
Chocolate Chip Cookie Dough ICE CREAM
Creamy vanilla ice cream with real cookie dough and chocolate chips
NATURAL AND ARTIFICIAL FLAVORS
ONE PINT (473mL)

KIRKLAND
Signature
100%
CAGE FREE
EGGS
Real
MAYONNAISE
64 FL OZ (2 QT) 1.9 L

KIRKLAND SIGNATURE

★ 53 ★

It's difficult to discuss Private Brands without including Costco's much-loved Kirkland Signature. The brand is the Private Brand portfolio at the warehouse club and has become synonymous with the retailer. It has redefined the nature of "brand-stretch," and despite the CPG naysayers who create narrowly defined category focused brands, Kirkland Signature has succeeded and grown to include hundreds of categories ranging from canned goods to premium wine to blue jeans.

The story begins in 1995, when Costco, unable to legally clear "Seattle Signature" as the brand name for their proposed private label, opted for "Kirkland Signature" instead, in honor of their flagship warehouse located in Kirkland, Washington.

Unlike traditional tier-focused grocery Private Brand strategies, Costco's Kirkland Signature is the brand, period. No tiers, no games, just a singular focus on quality, the customer and continuous improvement. There are no tiers of Kirkland Signature: no Kirkland Signature Ultra, Premium, Platinum, Gold or Select. The Kirkland Signature brand is a branded house of one. It is this focus, that has enabled the brand to develop a large group of brand loyalists who swear by all things Costco and Kirkland Signature.

So then why is it the 53rd brand?

Unfortunately, despite its success, Kirkland Signature has not applied its quality standards to its brand design, logo, and packaging. The brand is trapped in 1995, with a logo that anchors it to a long-gone private label era, a design strategy that is sporadic at best and fails to unify the brand in any way,or live up to the quality of the products in the package.

There is no question as to whether Kirkland Signature is a successful Brand — it absolutely is. However, it desperately needs a redesign and new logo. The redesign must create a cohesive, compelling design strategy that reinforces and builds on the inherent quality of the Kirkland Signature products. If Kirkland Signature combined the power of design with their heritage of great products, then they have the potential to raise the bar for the entire Private Brand industry and discover new heights of financial success and customer loyalty.

"The crowning achievement in my career
occurred when I had my epiphany to break my
addiction to running ½-price ads for Tide.
I realized that trying to attract customers
with national brands is a fool's errand.
I was going to break away from the pack.

Ultimately, the way I accomplished this was
with President's Choice."

Dave Nichol
President, Loblaws Companies - 1986

FIVE QUESTIONS

To close the book, I sat down with Fifty2 photographer Teri Campbell and Rob Wallace of Wallace Church to get their unique perspective on Private Brand.

Teri Campbell is the creative lead at Teri Studios and author of Food Photography & Lighting: a Commercial Photographer's Guide to Creating Irresistible Images.

After attending the Ohio Institute of Photography, Teri began his career as a photographer for P&G's in-house creative group. In 1988, he opened Teri Studios, a commercial photography studio specializing in food. His passion for food, combined with his artistic and business sensibilities, has uniquely positioned him to create appetite-appealing, proprietary imagery for consumer packaged goods companies and restaurants throughout the country. Clients include Long John Silvers, Bob Evans Restaurants, T. Marzetti, SuperValu, Kellogg's, Hershey's, and KFC among others.

Teri is a member of the American Society of Media Photographers (ASMP) and the International Association of Culinary Professionals (IACP). He also sits on the advisory board for Antonelli College.

When not in the studio, Teri is a frequent speaker at industry conferences and events like the International Conference on Food Styling & Photography, PDN's PhotoPlus Expo and the FoodPhoto Festival in Tarragona, Spain.

www.teristudios.com

What makes a great Private Brand?

A great Private Brand is one the consumer doesn't know is private. It's a brand that so completely engages the consumer he or she is unaware – or doesn't care – it's not a national brand.

How can a brand's visual identity help it differentiate in the marketplace?

By creating imagery that is unique, engaging, and specific to the target audience, retailers can show themselves as innovative and relevant, instead of just following the leader.

How does a retailer's visual brand identity help bring its brand to life?

In the age of Pinterest, where consumers are driven by visual messages, it is more important than ever for your brand to have a visual identity that connects emotionally with the consumer.

Of the Fifty2 brands in this book, what's your favorite and why?

Without question – Via Roma. I have loved that brand, since I first saw it at Design Annual. The idea that you would send a photographer to Italy to shoot the people featured on this package is inspiring. Not something I would expect from a small brand without a "national budget." But then again, I wouldn't expect that kind of inspired thinking to make it through the conference room mentality of many large corporations today. The images themselves are beautiful, authentic, engaging and ownable. A perfect example of what photography can do to set your brand apart.

What advice do you have for how retailers should work with photographers?

Understand that although much of the direction may have already been determined by the agency, the creative process should not end at the layout stage. Allow the photographer room to experiment and further your vision through the images they are creating. Sometimes retailers become so attached to a layout (that was meant only as a starting point), they don't allow the process to continue, thus confining them to mediocrity.

ROB WALLACE

★ MANAGING PARTNER, WALLACE CHURCH & CO ★

Rob Wallace, managing partner and strategic director of Wallace Church & Co, is responsible for strategic brand imagery action plans and consumer research. Charged with synthesizing marketing and design strategies, he ensures Wallace Church's design directions are focused on the most actionable and sales-effective goals.

Rob challenges himself to prove to the marketing community that corporate and brand identity is the most potent and cost-effective marketing tool.

What makes a great Private Brand?

Retailer-owned brands have come of age. No longer mimicking national brands or even following category conventions, great Private Brands anticipate and respond faster to changing consumer needs and expectations and challenge competitors to keep pace.

Great Private Brand strategy has surpassed the "as good as but cheaper" paradigm. Retailers have seen the return on their investment through increased product quality, which in turn has paved the way to delivering on a much more enticing promise.

Today, Private Brand marketers and their world-class strategy/design consultants only launch a brand if it embodies a truly compelling and differentiated brand experience.

How can design help retailers and their brands differentiate?

Great retailer-owned brand design lives at the very heart of the well-designed retail experience.

Not long ago, consumers complained that shopping was drudgery. The objective was to get in and out as fast as possible. Now, great retailers use a well-crafted mix of store design, merchandising, and Private Brands to engage and delight their customers. Customers no longer feel they are compromising when buying private brands. Great brand design rewards them. Retailers with well-designed store experiences report customers are spending more time in their stores, have more visits per week and have significantly increased their register rings. I'd argue their engagement is driven in large part by more effective retail brand design.

www.wallacechurch.com

Publix aluminum foil was designed more than 13 years ago and I believe it is as effective today as it ever was. The best design generates the highest return on investment, and these brands most certainly do.

How do you help RETAILERS select great design instead of obvious or easy design?

Just as the shopping experience has transformed from "drudgery to discovery," so has Private Brand design transformed from "the expected to the disruptive." We encourage our retail clients to make an investment in their Private Brand identities so that they drive a unique and engaging retail experience for their customers.

Great Private Brand design reinterprets category cues and resets expectations. At its best, great design is at the same time intuitive as well as somewhat surprising. It immediately and effectively communicates what the product is, but it goes well beyond "feature/benefit" expectations by shifting and exceeding expectations.

This "dance" between the expected and the disruptive is not only played out in the value-added or premium price tiers. Design of the most effective entry-price-point commodities also evokes a sense of delight. Look to Publix aluminum foil, for example. Can you imagine a more functional product? Look what Publix has done to shift that expectation. Just the simple twist of turning the product into a mini sculpture is inventive, distinctive and delightful. Retailers can look to this as inspiration in building their disruptive experiences.

Of the Fifty2 brands in this book, what's your favorite and why?

It's easy to admire iconic private brands like Craftsman, TUL and Kenmore that have transcended their original retailer and now become profit centers selling in additional channels. And, like all great brand architectures, Publix's Private Brand architecture just mentioned has also transcended time. Publix aluminum foil was designed more than 13 years ago, and it is as effective today as it ever was. The best design generates the highest return on investment, and these brands most certainly do.

But if I had to select a single favorite among the Fifty2, it would have to be Target's Archer Farms. Originally created by Michael Osborn Design, this articulate brand identity architecture continued to evolve as the brand hyper proliferated across literally hundreds of products and dozens of categories. Wallace Church & Co was proud to have been involved in recrafting the brand identity architecture allowing it to better segment across even more product categories. Now, effectively unifying hundreds upon hundreds of products, this identity is clearly a hallmark of one of the most prolific private brands of our time. As a result, Archer Farms has become a "destination brand " for the Target shopper, and I'd suggest, one of Target's most valuable assets.

What advice do you have for retailers working with design consultancies?

Embrace Apple's philosophy: in order to lead, you have to "think differently."

Build your relationships with your design consultancies around mutual trust and respect. Encourage them to challenge you, and remain open to change. Listen to them. Many work with national brands that invest millions in consumer research and innovation. While consultancies can never betray confidential information, they remain acutely aware of changing consumer needs, expectations and behaviors. Trust them, and allow them to prove out their recommendations.

> "Stay nimble. Stay courageous. Stay the renegade."
>
> Rob Wallace

Consider your brand strategy and design partners brand stewards. Engage them on a longer-term basis rather than project-by-project. Allow them to become fully immersed in your brand and corporate culture. Select those who can show a track record of evolving a brand architecture over time by anticipating market dynamics. Respect and learn from their processes and best practices, and adopt them as your own.

Work together to quantify the value of design, prove its ROI, and return some of that incremental value back into design. Fuel your new product development and package innovation budgets with a portion of the incremental value that design generates.

How can private brand owners avoid the mistakes of the past and the missteps of national brands?

I believe the successes of great Private Brand design are so evident and unequivocal that private branding will never revert back to its old copycat ways. Design has earned its respect within the smartest senior retail management. While it still has a long way to go to become a driver of retail corporate culture, I have little fear it will revert to the mistakes of the past.

What I do fear is that design is not appropriately funded with the proper time, resources and money needed to truly generate its fullest potential. I've too often heard, "We are not a national brand, and we don't have those kinds of budgets." This kind of old thinking needs to evolve. Once design has proven its paramount financial contribution to retail brand success, then appropriate budgets, timetables and best practices will be respected.

What I fear even more is the over indulgence of "success." I would argue that one of the reasons Private Brand design is effective is because it's managed by fewer leaders. As Private Brand management matures and grows, it's my fear that it will become overburdened with tiers of "decision makers." If so, it can easily fall back on the bad habits of some national brand practices, which over-analyze and often over-complicate the design process. More simply stated, identities that are "design by committee" rarely achieve greatness.
And so my advice?

Stay nimble. Stay Courageous. Stay Renegade.

ABOUT THE AUTHOR

CHRISTOPHER DURHAM

★ VICE PRESIDENT OF RETAIL BRANDS, THEORY HOUSE ★

Christopher Durham is the vice president of retail brands at Theory House, the branding and retail marketing agency . He is a consultant, strategist and retailer with close to 20 years of real-world retail and corporate experience creating, launching and building billion dollar Private Brands. His influential website, My Private Brand (www.mypbrand.com), seeks to drive the changing Private Brand landscape, focusing on the emerging art and science of Private Brand management. With readers from more than 67 countries and more than 3,000 stories, he has worked to push the traditional industry from the dark ages of private label into the new era of retailer-owned BRANDS.

While working at Lowe's Home Improvement, he developed and implemented the strategy for their multi-billion dollar portfolio of Private Brands, as well as creating and managing many of those brands. Prior to Lowe's, he served as brand manager at Delhaize America, where he developed and delivered retail brand marketing as well as Private Brand strategy and development for Food Lion, Bloom, and Bottom Dollar.

He is a trusted consultant for many of the world's leading retailers, including Office Depot, Family Dollar, Best Buy, Lowe's Home Improvement and Grainger.

Durham is the co-author of the groundbreaking in-depth analysis of Walmart's private brand portfolio, *MPB SIGHTLINE: The 2013 Walmart Private Brand Portfolio*. He is a much sought-after thought leader who has been featured in columns and interviews in the *Washington Post*, *Brand Packaging*, *Private Label Buyer*, *Global Retail Brands*, *Food Processing*, *Retail Leader*, *PLMALive!* and *Food Processing Magazine*.

Dynamic in his presentation while down to earth and frank in his opinions, he has presented at numerous conferences, including FUSE, The Dieline Conference, Packaging that Sells, Private Brand Movement, Shopper Insights in Action, FMI Private Brand Summit, Private Label Buyer Conference and the Own Label Conference in London, England.

He lives in Charlotte, NC, with his wife, Laraine, and daughters, Olivia and Sarah.

mypbrand.com ★ theoryhouse.com

RETAILER INDEX

"You cannot buy the revolution.
You cannot make the revolution.
You can only be the revolution.
It is in your spirit, or it is nowhere."

Ursula K. Le Guin

CPSIA information can be obtained
at www.ICGtesting.com
Printed in the USA
LVIC05n1357240414
383103LV00011B/25